ROBOTS

ROBOTS

ROBOTS

Edited by

Harry M. Geduld   Ronald Gottesman

New York Graphic Society   Boston

Copyright © 1978 by Harry M. Geduld and
Ronald Gottesman
"Genesis II: The Evolution of Synthetic Man"
Copyright © 1978 by Harry M. Geduld
English translation of "The Anguish of the
Machines" Copyright © 1978 by Anna Lawton

First Edition

The lines by David McCord on page 185 are taken
from "The Ballade of Time and Space," which
appeared in *Bay Window Ballads* (New York:
Charles Scribner's Sons, 1935). Reprinted by
permission of the author.

Library of Congress Cataloging in Publication Data

Main entry under title:

Robots, robots, robots.

    Bibliography: p.
    1.   Automata—Literary collections.   I.   Geduld,
Harry M.   II.   Gottesman, Ronald.
PN6071.A94R6   808.8'0356   78-1435
ISBN 0-8212-0688-5

Designed by Janis Capone

Picture Research by Sally Fox

*New York Graphic Society books are published by
Little, Brown and Company.*

*Published simultaneously in Canada by Little,
Brown and Company (Canada) Limited.*

PRINTED IN THE UNITED STATES OF
AMERICA

Automatically
Dedicated to our favorite humans —
Daniel, Grant, Lann, and Marcus —
With our love

DOMIN. Robots are not people. Mechanically they are more *perfect* than we are; they have an enormously developed intelligence, but they have no soul.

HELENA. How do you know they have no soul?

DOMIN. Have you ever seen what a Robot looks like inside?

— Karel Čapek, *R.U.R.*

ROTWANG. At last my work is ready. I have created a machine in the image of man, that never tires or makes a mistake. Now we have no further use for living workers. . . . Give me another 24 hours and I'll give you a machine which no one will be able to tell from a human being.

— Fritz Lang, *Metropolis*

Contents

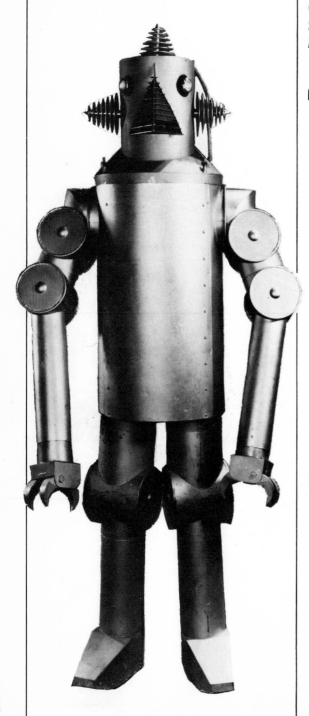

ROBOTS

ROBOTS

ROBOTS

Genesis II: The Evolution of Synthetic Man

by Harry M. Geduld

Robot . . . [Czech, f. robota statute labour, robotnik serf.] Applied by Karel Čapek in his play R.U.R. to a mechanical apparatus doing the work of a man; hence, any such apparatus; a person whose work or activities are entirely mechanical; an automaton. . . . Hence Robotesque a., resembling or suggestive of a robot. Robotian a., of or belonging to robots. Robotism, mechanical behavior or character. Robotization, the process of making or fact of being mechanical in behavior. Robotize, v. trans., to subject to the operation of robots; to render mechanical. . . . Robotry, the business, position or condition of robots.

— Oxford English Dictionary

Myth

In the beginning, before the machine, there was myth. All myths about synthetic and automatic human beings originated in the same fundamental beliefs: the almost universal convictions that man was the sublime and ultimate creation of the gods (or God) and that he was made in the image of the Divine. The latter notion was a variation on the widespread mythic conception of anthropomorphic gods.

These beliefs are, of course, formally expressed in Genesis together with one of the earliest and best-known accounts of the creation of man. We are told that Adam was made out of the dust of the earth, and the words of the burial service remind us that he returns ultimately to the common clay whence he came.[1] The story indicates that God has made man as part of

Nature and not as something superior to it. Thus, by implication, it would be a violation of Nature for man to be made without the hand or sanction of God. Furthermore, as the creation of man was the capstone of divine achievement, it would be sacrilegious or worse — a disastrous act of hubris — for any human being to attempt to emulate the Creator by making intelligent life.

The early myths of many races affirm that the gods have the power and sole prerogative to create human life out of anything they please — though they may not always be satisfied with the results from either a physical or an ethical standpoint. Thus the Popol Vuh, scriptures of the Guatemalan Indians, recount three successive attempts of the gods to create men.[2] The first effort was a total failure. Like Jehovah, the gods

---

[1] The biblical story is derived from a Mesopotamian myth in *The Epic of Gilgamesh* (c. 2000 B.C.): "Araru, the goddess of creation . . . dipped her hands in water and pinched off clay; she let it fall in the wilderness, and `noble Enkidu was created."

[2] The Popol Vuh cannot be dated exactly but belongs to the pre-Columbian era. The text was first written down by the Spanish monk Francisco Ximenez during the seventeenth century. Ximenez preserved the work in the original Quique language.

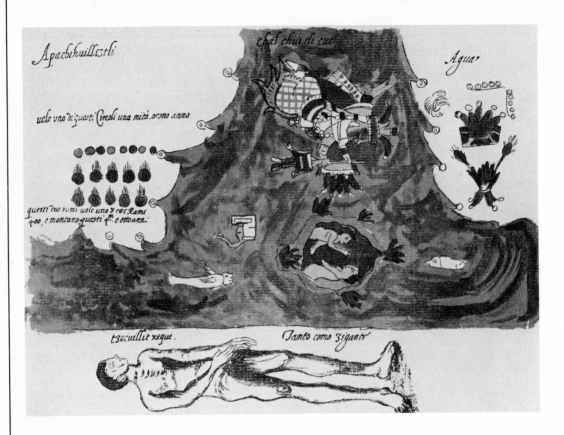

In myth, almost universally, man is one of the first synthetic creations.

The ancient Egyptians believed that the god Neph shaped the first human beings on his potter's wheel. Hou Chi, the founder of the first Chinese dynasty, was said to have been born from a footprint of Mu King, god of the immortals. According to Persian lore, Ahura Mazda, Lord of Wisdom, formed the first youth out of the sweat of his body. Genesis tells us that God created "man of the dust of the ground and breathed into his nostrils the breath of life; and man became a living soul."

Shown here is the Aztec creation myth as illustrated in the Codice Rios Messicano Vaticano 3738: "The earth was without men. And the great god Quetzalcoatl brought bones from the underworld. And lo, he stumbled as he walked upon the earth, and the bones fell from his grasp. They were seized by quails who gnawed upon them. Then came the god Ciuacoatl who took them from the quails and ground them into paste. And from this paste he fashioned the first men...." (Peabody Museum, Harvard University. Photograph by Hillel Burger)

molded men out of clay. But these creations remained inanimate and were eventually washed away by the rains. Next they shaped puppetlike forms out of wood. These could be made to move, but were still unsatisfactory. So the gods destroyed most of them. The few that escaped destruction became the first monkeys. At last, the gods made men of flesh and blood, who could worship their creators. And with these they were satisfied.

But Jehovah was less easy to please. Looking over His Sixth Day's work, He saw that it was good — but only physically. The rest of the Bible provides abundant evidence of God's dissatisfaction. (One of the earliest post-Edenic stories is that of the Tower of Babel. This myth of man's attempt to challenge God prefigures many golem/robot stories in which man's creation challenges *its* creator.) And surely, if God is discontented with the human beings He created, how much more objectionable to Him would be a synthetic man created by men?

4

The formative or prototypic stories that we shall now consider must be viewed against this background of ancient beliefs. Prior to the twentieth century, there were three distinct traditions concerning the creation of artificial beings: first, a body of mythic or legendary material about god-sanctioned creation; second, a contrasting corpus of narrative about sacrilegious creation; and third, the history of *real* automata, forming a chapter in the development of technology rather than another aspect of mythology. Both traditions of mythic material involve man's relationship with his *creator*; the third tradition, however, focuses mainly on man's relationship with *his* most consequential *creation* — the machine, and this in turn shades into the Mechanist vs. Vitalist controversy concerning whether man is himself a machine.

From Galatea to the Golem: God-Sanctioned Creation

The Greek story of Pygmalion and Galatea typifies the tradition of artificial creation by divine sanction. Pygmalion, a mere mortal, falls in love with the goddess Aphrodite. When the goddess refuses to accept him as her lover, Pygmalion makes an ivory statue of Aphrodite and takes it to bed with him. Touched by his adoration, the goddess brings the statue to life as the maiden Galatea, who becomes Pygmalion's bride.[3]

Several comments are worth making about this famous myth. Although it recounts a story of artificial creation, no sacrilege, no act of hubris is involved. Galatea is brought to life through divine power and pity: it is the goddess Aphrodite not the mortal Pygmalion who actually inspires and then animates the statue. True, Pygmalion provides the form — but even that was made in the image of the goddess. In effect, the creation of Galatea is simply a reward for piety; in no way does it demonstrate either man's technological ingenuity or his rivalry with the gods.

In another legend, the goddess Athene directs the mortal Cadmus to sow the teeth of a dragon he has killed. From these teeth spring an army of warriors whom Cadmus tricks into fighting one another until only five remain. With these survivors he builds Cadmea, the citadel of Thebes. Subsequently, the five dragon's-teeth warriors become the nonhuman ancestors of the Theban aristocracy.

The giant Talos, another animated statue of Greek myth, was also a divine creation, but unlike Galatea or the warriors of Cadmus, its purpose was inimical — specifically to Crete's enemies. Aphrodite's husband, Hephaestus, the Greek god of fire, fashioned the monstrous figure out of bronze and gave it to King Minos to serve as a guardian of the approaches to Crete.

Hephaestus seems to have had a special interest in, and talent for, making animated figures. In addition to Talos, he is said to have made a number of articulate, mechanical females out of gold and used them to do much of the work in his smithy. He also constructed several three-legged tables that were capable of moving about by their own volition.

But Talos appears to have been Hephaestus's crowning achievement. The statue was virtually invulnerable: its only weakness was the equivalent of a tiny blood vessel in the leg. Until the arrival of Jason and the Argonauts, Talos kept Crete free of strangers. When anyone landed on the island, the statue's body would become glowing hot and the stranger would be grabbed by Talos and then crushed and burned against the statue's body. But through Medea's magic

---

[3] Bernard Shaw "recapitulates" the story in Part V of *Back to Methuselah*, but adds the "Frankenstein" touch of having his Pygmalion destroyed by one of his creations.

arts and her knowledge of Talos's "Achilles' heel," the statue was destroyed and the Argonauts were able to land in Crete.

Daedalus, a legendary descendant of Hephaestus, is said to have made statues that could move by themselves: one of them was a bronze warrior that also resisted the Argonauts; another was a wooden figure of Aphrodite that was made to move by the use of quicksilver.

There are some significant similarities and differences between the stories of Hephaestus and Daedalus and the creation of Galatea. These tales — like the Popol Vuh and the Mesopotamian and biblical creation myths — reinforce the idea that the gods have the power to make man — or at least animated figures — out of anything they choose: clay, ivory, bronze, wood, dragon's teeth, etc. Second, both Galatea and Talos were made in the image of their creators — although this may be less evident in the case of the latter than with the former.[4] However, the contrasts between the two myths are, perhaps, more important than the similarities. Galatea was a goddess's gift to man, whereas all the moving figures of Hephaestus were made to serve the gods or their children (Minos was a son of Zeus). And where Aphrodite's animated statue became the consort of a mortal, Talos was a god's device for destroying men. Finally, the fundamental import of the two myths is quite different. The Pygmalion and Galatea story concerns the relationship between love and artistic inspiration; the tale of Talos and Medea, on the other hand, seems to deal mythologically with the overthrow of Bronze Age civilization. In neither story does the creation or destruction of synthetic figures involve a direct rivalry between man and the gods.

Talos and Galatea are the most famous of the animated statues of classical myth.

Too numerous to discuss here are the many other ancient and medieval tales of statues brought to life. But it is worth noting that philosopher-scientist Empedocles (fifth century B.C.) was perhaps the first *mortal* reputed to have animated a statue: in his case, a figure of bronze. (We shall have more to say later about Empedocles — in connection with the philosophical debate over Mechanism.) A Sicilian legend of the fifth century A.D.[5] refers to a statue which drove away all the enemies of the Sicilians and also subdued the fires of Etna. And in the Middle Ages, one of the many popular stories about the "magical powers" of Virgil attributed to the poet the construction of a marvelous statue that dispersed sulfurous blasts of hot air from Vesuvius.[6]

After the myth of Galatea, the most famous and enduring stories of God-sanctioned synthetic creation are the tales of the golem. In its most familiar form, the golem legend tells how, in the sixteenth century, Rabbi Loew of Prague saved his people from an almost inevitable pogrom by invoking God's assistance in the making of an animated clay giant who became the protector of the Jews.

Rabbi Judah Loew Ben Bezalel was a real person. An authoritative Kabbalist and mystic, known popularly as the *Maharal*, he lived in Eastern and Central Europe circa 1525–1609. His gravestone can still be seen in the old Jewish cemetery in Prague.

The golem stories associated with Rabbi Loew's name actually represent the legend in a comparatively late phase. It will be enlightening here to look back and consider the origins of the legend before turning to its later significance and influence.

Golem is a Hebrew word meaning *formless mass*. The word occurs once in the Bible, in Psalms 139:16, where it alludes to the human embryo or fetus. It also seems to have

---

[4] Actually both Talos and Hephaestus had weaknesses of the legs, and the glowing heat of the statue's body suggested the fire god at work in his smithy.

[5] Olympiodorus in *Photius*, cod. 90.

[6] See further Domenico Comparetti, *Vergil in the Middle Ages* (London: Allen & Unwin, 1966).

become a term employed by Gnostic and Neo-Platonist rabbis to signify undifferentiated matter or "substance" — without form or vitality. Other, less philosophical, rabbis, some of whom were also authors of Talmudic commentaries, used the word in imaginative conjectures about the origins of life in general and the creation of Adam in particular. Among these rabbis was the circle of Jahanan Ben Zakkai (circa second century A.D.) whose preoccupations with the mysteries of Creation included the study of the power of God's secret name: the *Shem-Ha-Meforash* or Tetragrammaton, sometimes cryptically represented by the letters YHVH. In medieval Jewish legend, *Shem-Ha-Meforash* was to become the chief magic formula for bringing the golem to life. Other formulas were to be derived from the *Sepher Yetzirah* or *Book of Creation*, composed circa the third century A.D., a Hebrew work of letter-mysticism, regarded by many Kabbalists as an authoritative treatise on the secrets of the cosmos.

At this early period of the Christian era, the first legends of the golem became absorbed into Talmudic commentaries. For example, included in the Haggadah or non-legal segments of the Talmud is the story of the learned Rabbah Bar Nahmani (third century A.D.) who created a golem. The creature was dumb, but in all other respects was just like an ordinary human being.[7] The nonhuman nature of Rabbah's golem was discovered by Zera, another rabbi, who instantly reduced the creature to lifeless clay and then denounced Rabbah for having committed sacrilege.

Aside from such early legends, the Talmud also contains fairly explicit instructions for creating a golem:[8]

How was Adam created? In the first hour his dust was collected; in the second his form was created; in the third he became a shapeless mass [Golem]; in the fourth his members were joined; in the fifth his apertures were opened; in the sixth he received his soul; in the seventh he stood up on his feet. . . .

Eventually, two distinct conceptions of the golem existed among medieval Jews. In Germany, rabbinic students of the Kabbala sometimes referred to "making the golem" as a metaphor for a state of religious ecstasy that was known to have been experienced after the prolonged practicing of sacred rituals. This idea was, however, gradually submerged beneath the more popular conception of the golem as an actual artificial figure made to serve its creator and to protect the Jewish people in their hour of peril.

The legend of Solomon ibn Gabirol of Valencia (A.D. 1021–1058) preserves the tradition of the golem as servant. This particular rabbi created a female golem to do his housework. When the king heard about it, he gave orders for Rabbi Solomon to be executed for being a sorcerer. Fortunately, the good rabbi was able to save himself by showing the king that he could undo his magical work quite easily by turning the golem into a heap of dust. In the following century, another rabbi, Samuel, father of the celebrated Judah Chassid, was said to have had a male golem as his bodyguard. In keeping with the tradition, neither this golem nor the earlier, female one was able to talk.

Belief in golem-servants was widespread among Jewish communities at least through the eighteenth century. And Gentiles often knew about them even if they did not always believe in their existence. One who did accept them as fact was a German anti-Semite, Johann Jakob Schudt, who, in his *Jüdische Merkwürdigkeiten* (*Jewish Wonders*) of 1718, mentioned that "Polish Jews often make the golem, which they employ in their homes . . . for all sorts of housework."

The many non-Jewish analogues to the

---

[7] The notion that the golem lacked the power of speech was to become a persistent element of the legend; it relates to the traditional Jewish belief that the ability to talk was a gift that could be granted only by God.

[8] See N. Ausubel, *The Book of Jewish Knowledge* (New York: Crown Publishers, 1964), p. 186.

A *homunculus is created by Wagner, Faust's assistant. "Homunculi quanti sunt!" ("What dwarfs men are!") says Plautus in his Captivi. But the homunculi attributed to the medieval alchemists, to German necromancer Georg Faust (1480?–1538?), and to Swiss physician Bombastus Palacelsus (1493?–1541) were not dwarfs but miniature humanoids chemically synthesized in alembics or glass retorts and imaginatively evoked centuries later in the pages of Goethe's Faust. (The Bettmann Archive, Inc.)*

golem-servant are too numerous to discuss here. They include examples as varied as the aforementioned statue which Virgil brought to life and Goethe's celebrated poem of the Sorcerer's Apprentice. In many of these accounts, as with the later versions of the golem legend, earlier folkloric traditions became inextricably mingled with medieval preoccupations with alchemy. That is, white and black magic became confused. Perhaps inevitably: for if devils like Faustus's Mephistophilis (in Marlowe's play) could be raised by incantations and alembic concoctions, why not non-Satanic beings like Faustian homunculi or golems? Hence, between the primitive science of Paracelsus (whose experiments were said to have included the manufacture of homunculi) and

the "magic" of Rabbis Elijah of Chelm and Loew of Prague (who made the most famous golems) the distinction was negligible. With the former no less than with the latter, superstition and mystical formulae were integral to the creation process. Paracelsus (1493?–1541) is sometimes regarded as a seminal figure, a pioneer of modern science. But not for his making of homunculi. That was essentially nonscientific, and in that sphere of activity the work and influence of the rabbis certainly outstripped him.

In the legends of Rabbis Elijah and Loew, the golem was to become something more than a mere servant or companion-body-guard. At first a champion, he gradually assumes the role of a potential menace — a would-be destroyer of his creator and perhaps of all men. We are almost within sight of *Frankenstein.*

Rabbi Elijah Ben Judah of Chelm (1514–1583) was renowned as a healer, but the monster he created became even more famous for its powers of destruction. Here, perhaps, began the tradition of the golem/robot's creator losing control of his creation — an obvious analogy to Adam's original disobedience to *his* Creator. Rabbi Elijah brought his golem to life by molding it out of clay and pasting onto its forehead the Kabbalistic formula, *Shem-Ha-Meforash.* The monster's live aspect instantly repelled the rabbi, but he was even more aghast to discover that his golem reveled in destruction. Whereupon, before losing total control of the creature, Rabbi Elijah plucked the formula from its forehead, and the golem immediately crumbled into dust.

The stories of the golem of Rabbi Loew are obvious elaborations of the tales of Rabbi Elijah; they also provide the legend in its most developed form. First published in the seventeenth century in the *Nifluot Maharal (The Maharal's Miracles),* the tales of Rabbi Loew's golem became widely known in the twentieth century through Chaim Bloch's charming retelling: *The Golem: Mystical Tales from the Ghetto of Prague.* Either the original *Nifluot Maharal* or Bloch's *The Golem* inspired writers as varied as Rilke and Kafka, Mary Shelley and the Brothers Grimm, and works as varied as Gustav Meyrink's novel, *The Golem*, and H. Leivick's[9] memorable Yiddish verse drama of the same title, and many golem films, including three notable movie versions by German director Paul Wegener.

The golem legends associated with Rabbi Loew began in the year 1580, when a fanatical priest named Thaddeus accused the Jews of Prague of murdering Christian children and using their blood in making Passover matzoh. After trying without success to reason with Thaddeus and his followers, Rabbi Loew realized that the accusation of ritual murder was the prelude to a pogrom. In a dream, he sought and received advice from Heaven — to make a golem that would confound Israel's enemies. Upon awakening, he requested his son-in-law and one of his pupils to join him beside the river Moldau. There — with their assistance — the golem was created. It was like a human being in all respects except for its great height, its superhuman strength, and its inability to speak.[10]

After dressing his creation in the garb of a *shammes* (sexton), Rabbi Loew spoke to it thus: "Know that we have created you out of common clay. Your work is to protect the Jews from their enemies. Your name is Joseph [Yosele] and you will serve me and live in my house. You must obey my commands whatever they may be — even if I order you to jump into a furnace or leap into the ocean." The golem was dumb but he could understand the rabbi's words. However, because he was mute and seemed stupid, the people called him *Chomer*

H. Leivick's Yiddish play The Golem, *which premiered in 1925, dramatizes the most enduring medieval legend of God-sanctioned synthetic creation. Here is an electrifying scene from act three of the Habimah National Theatre production, starring Aharon Meskin as the Golem and Ada Tal as Debora, directed by B. Varshilov. (Habimah National Theatre, Tel Aviv)*

(Idiot) *Golem*, and among the ghetto Jews the word golem by itself gradually became synonymous with "idiot."

Chaim Bloch's book recounts many subsequent adventures of the rabbi and his creation — too many to summarize here. But, in brief, the golem carried out the tasks for which it had been made. It patrolled the ghetto streets and terrorized the accusers and would-be destroyers of the Jews. It also undertook the rabbi's household chores and became the ghetto's hewer of wood and drawer of water. But, at last, the rabbi put an end to it.

There are three distinct versions of why this occurred. The first is simply that, after the Jews had been left in peace for some years, it was decided that the golem was no longer needed: this may be interpreted as a conciliatory gesture toward the non-Jewish community of Bohemia. A second version is that the creature stupidly and sinfully

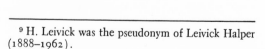

[9] H. Leivick was the pseudonym of Leivick Halper (1888–1962).

[10] Some modern commentators consider Rabbi Loew's golem to be the first example of an *android* —a robot with human (rather than mechanical) appearance and characterized by some of the worst or most primitive of human passions.

The golem legend has endured in the form of many screen adaptations. Shown here is a poster for Julien Duvivier's version of 1935. (American Jewish Historical Society)

continued its manual labors on the Sabbath when the rabbi could do nothing to control it. (In modern terms, the golem behaved in this way because it lacked any "feedback" device.) A third version, obviously prefiguring *Frankenstein*, is that it revolted and tried to dominate its creator. Realizing that the golem had become an evil monster with a lust for power and destruction that threatened the existence of all the Jews, Rabbi Loew and his two companions waited until it was asleep, and then destroyed it by going through the creation rituals in reverse order. In a variant story, the golem was running amok in the ghetto, when the rabbi confronted it and ordered it to take its boots off. Automatically, the monster sat down to obey the command. Its head was then low enough for the rabbi to pluck from its mouth the parchment inscribed with the *Shem-Ha-Meforash*. This promptly immobilized — or demolished — the creature.

Notwithstanding the third of the variant endings, the legend's emphasis is not on the perils or penalties of creating an artificial man, but on the Jewish wish for a guardian — as distinct from a Messiah — in the hour of oppression. (It is not impossible that such real-life rabbis as Elijah of Chelm and Loew of Prague actually concocted or spread tales about golems in order to terrify anti-Semites.) It is also noteworthy that there is no version of the golem legend in which the creation of the monster is referred to as a transgression of divine right. The creature was God's work as surely as Galatea was Aphrodite's. The rabbis acted merely as God's agents. In every instance, making the golem was actually a deed of piety, and the symbolic involvement of the four elements in the creative process demonstrated that it was not an unnatural experiment but a harmonious interaction of the powers of Nature and the powers of God. Early versions of the legend in which "making the golem" was synonymous with ecstatic experience established the association between the creative process and communion with God. As the making of Adam was the Creator's most direct and intimate involvement with man, so the ritual of making the golem became, to the mystics, a kind of sacramental reenactment of the primal involvement of God and man. God's very breath had given life to humanity — the antithesis of the sacrilegious guilt we associate with the creative act of a Frankenstein.

The intermediary concept of the golem as servant, while less profound than the golem as mystical ecstasy, was essentially a pious notion. Daily toil minimized the time that man could spend in worshipping God. The golem-servant, by taking over the arduous work of the ghetto Jews, liberated them for more religious activity. The golem-servant was, of course, a precursor of the robot-servant of innumerable science fiction stories and movies. It also had another significance. It was an artificial equivalent of a housewife. The golem was "created" by men with the aid of a "masculine" God, and became the substitute for a female. This idea of male creativity and self-sufficiency has been recurrent in science fiction through *2001: A Space Odyssey*.

The tradition of the golem as champion of the Jews, while obviously of more secular import than the other traditions, nevertheless retained traces of the biblical Covenants: through the golem, God was continuing to insure that His people would survive. The golem's success against the enemies of the Jews was an affirmation of God's power and a demonstration of His reward for the piety of those who truly believed in Him.

From Prometheus to Frankenstein: Man's Rivalry with God

In contrast to the traditions of God-sanctioned artificial creation, a second body of myth and narrative, frequently confused with the first, deals with artificial creation as sacrilegious disobedience. The primal myth here is the story of Prometheus.

Our knowledge of this myth is derived from various classical sources which provide different emphases and sometimes different details. We should, first of all, distinguish the Greek Prometheus from the Roman Prometheus. Roughly speaking, the Greek stories concern Prometheus *pyrphoros* (Prometheus as bringer of fire), while the Roman tales concentrate on Prometheus *plasticator* (Prometheus as creator or re-creator of man.)[11]

In the ancient Greek Hesiodic poems, *Works and Days* and *Theogony*, "wily Prometheus," the son of the Titan Iapetus and the Oceanid Clymene, is an archtrickster who attempts to deceive Zeus by offering him what looks like the meat of an ox; it is actually some bones wrapped in fat. Zeus sees through the deception and decides to punish Prometheus by depriving mankind — whom Prometheus loves — of the gift of fire. Not to be outdone even by Zeus, Prometheus steals fire — concealed in a fennel stalk — and gives it to the mortals. When Zeus discovers what has happened this time, he determines to punish both the thief and those who received the stolen gift. So, for thirty thousand years, Prometheus is bound to a rock in the Scythian desert. By day an eagle swoops down and tears out his liver, which grows back during the night. Man's punishment, more insidious than that of his benefactor, is the creation of woman — regarded by Hesiod as a "thing of evil." At Zeus's direction, Hephaestus molds the first woman out of clay. She is called Pandora. Then, dressed and beautified by Athene, Pandora is revealed to all the gods and men: "And wonder took hold of the deathless gods and mortal men," says Hesiod, "when they saw that which was sheer guile, not to be withstood by men. For from her is come the deadly race and tribe of women who live among mortal men to their great trouble."

[11] This distinction is derived from M. K. Joseph's Introduction to his edition of *Frankenstein*, London: Oxford University Press, 1971.

A Graeco-Roman gem that depicts the Greek myth of the creation of man as recounted by Ovid in his Metamorphoses. *The Titan Prometheus, son of Iapetus, used clay and water to fashion men in the image of the gods. When his work was completed, the goddess Athene breathed life into his creations. The torment and triumph of Prometheus — he was punished for stealing Olympian fire for man — was sung by Shelley in* Prometheus Unbound. *Beethoven, in his ballet* The Creatures of Prometheus, *celebrated the beings for whom the Titan had plundered the treasure of heaven. (Reproduced by courtesy of the Trustees of the British Museum. Photograph © Werner Forman Archive)*

Other sources, however, are far more significant. Plato tells a tale different from Hesiod's. In his Socratic dialogue *Protagoras*, we learn that the gods fashioned mortal creatures out of a mixture of earth and fire and then directed Prometheus and his brother Epimetheus to endow them with suitable qualities. Epimetheus told Prometheus to leave the work to him, but then proceeded to distribute all the qualities he possessed (brute force, protective hides, flight, etc.) to the animals and birds. When he came to man, he had nothing left to give. Prometheus, concerned that the human race should not be left naked and helpless, proceeded to compensate for his brother's oversight by stealing fire and the mechanical arts of invention from Hephaestus and Athene. These he bestowed on man, but his crime was discovered and he was punished — essentially, as Plato points out, for his brother's blunder.

As we noted earlier, the Roman traditions of Prometheus were at variance with the Greek myths. Ovid, writing several centuries later than Plato, preserves two alternative accounts of man's creation: first, that Zeus "made man of his own divine substance"; and second that Prometheus, "out of a mixture of earth and water, molded the body of the first human being in the form of the all-controlling gods" (*Metamorphoses*, Book I). Man's soul was derived from vestiges of ethereal elements that remained from the original creation of the Universe. As Robert Graves has noted, the Promethean creation story — like the one in Genesis — was rooted in the *Epic of Gilgamesh*. About a century after Ovid, Lucian (A.D. 115–200) wrote a comic playlet — *Prometheus on Caucasus* — in which Hermes addresses Prometheus while the latter is being crucified by Hephaestus. Hermes' comments embody material from both Hesiod and Ovid, clearly indicating that the story of Prometheus as man's creator had become part of the accepted mythic tradition:

Prometheus . . . you . . . cheated Zeus by wrapping bones in glistening fat. . . . Then you made human beings, thoroughly unprincipled creatures, particularly the women; and to top all, you stole fire, the most valued possession of the gods, and actually gave that to men.

The Literary Prometheus

Roughly contemporaneous with Lucian, Pausanius, in his *Description of Greece*, endeavored to provide factual support for belief in the Promethean creation story by specifying where it had occurred and what vestiges remained:

At Panopeus there is by the roadside a small building of unburnt brick, in which is an image of Pentelic marble, said by some to be Asclepius [god of medicine], by others Prometheus. The latter produce evidence of their contention. In the ravine there lie two stones, each of which is big enough to fill a cart. They have the color of clay, not earthy clay, but such as would be found in a ravine or sandy torrent, and they smell very like the skin of a man. They say that these are the remains of the clay out of which the whole race of mankind was fashioned by Prometheus.

A curious addition to the Prometheus myth is found in the *Bibliotheke* or *Library*, an anonymous book often erroneously attributed to Apollodorus of Athens (circa second century B.C.). In this work, which has obvious analogy to the story of Noah, it is not Prometheus but his son who is involved in the creation of man. Deucalion, son of Prometheus, married his cousin Pyrrha, daughter of Epimetheus. It was the Bronze Age and men already existed, but Zeus, angry at their possession of fire, resolved to drown them in a great flood. Most were, indeed, destroyed in the deluge, but a few survived by fleeing into the mountains. Deucalion and Pyrrha, having been warned — by Prometheus — of the impending disaster, took refuge in a large chest which they had fitted out with provisions. After floating about for nine days and nine nights, they landed on the slopes of Parnassus. In gratitude for his survival, Deucalion made sacrifice to Zeus — who was god of escapes. Zeus, responding to the sacrifice, sent Hermes to reward Deucalion with anything he desired. Deucalion asked for the creation of new men. Zeus thereupon directed him and his wife to throw stones over their heads. Those thrown by Deucalion became men; those thrown by Pyrrha became women. Hence, explains the author of the *Bibliotheke*, "people were called metaphorically people (*laos*) from *laas*, a *stone*." (The "equation" of people with stones seems analogous to the traditions of creation from statues and creation from clay.)

Of the two main traditions — Prometheus *pyrphoros* and Prometheus *plasticator* — it is, patently, the latter that most concerns us here. The *pyrphoros* has gradually earned the reputation of suffering benefactor of mankind, more or less obliterating his earlier (Hesiodic) reputation for cunning and deception. The exaltation of Prometheus was largely the work of Aeschylus (*Prometheus Bound*) and the Romantic poets (particularly Byron and Shelley). To these writers he appeared both as the original culture-hero who brought or taught the arts of life to the human race, and (primarily to the Romantics) as the archetypal liberal hero in revolt against reactionary tyranny. If the *pyrphoros* became Christ-like, the *plasticator* increasingly assumed a role close to the Satanic. To steal and to suffer for mankind was sublime, for Prometheus's gift of fire symbolized the potential of human enlightenment; but his making of men came, in due course, to be interpreted as sacrilege or hubris.

Prometheus was sometimes referred to as *Demiurge*, that is, as secondary deity or creator. Zeus had made the first men, but if *he* was unable to perfect them, how could a secondary deity expect to improve on them? Roman writers such as Lucian made no bones about the fact that Prometheus was, at least in part, being punished for re-creating something as evil as the human race. It is not surprising that among the Gnostics and other early philosophers, Demiurge became a term synonymous with the Origin of Evil.

In one important respect, however, Prometheus as teacher of mankind became identified with Prometheus as creator: the teacher inspired not only the arts of life but also the urge to create. In *The Moralists* (1709), the Earl of Shaftesbury considered whether Prometheus was "a Name for Chance, Destiny, a Plastic Nature, or an Evil Daemon [the last expression prefigures the name given to Frankenstein's monster]. . . ." He then went on to discuss the creative impulse: "We have a strange Fancy to be Creators, a violent Desire at least to know the Knack or Secret by which Nature does all. The rest of our Philosophers only aims at that in Speculation, which our Alchymists aspire to in Practice. For with some of these it has been actually under deliberation how to make *Man*, by other Mediums than Nature has hitherto provided." M. K. Joseph has speculated on whether Mary Shelley read this passage prior to writing *Frankenstein* (1818). Whether she did or not, she certainly expected her readers to draw the appropriate mythic implications when she subtitled her novel, "The Modern Prometheus."

The story of *Frankenstein* is too well known to require summarizing, but its contrast and relationship to the foregoing myths and legends are less familiar.

M. K. Joseph observes that "Mary Shelley . . . alone [of all the Romantics] seized on the vital significance of making Prometheus the creator rather than, as in Byron and Shelley, the suffering champion of mankind." To this perceptive comment we should add that the "Modern Prometheus" — who is human (Victor Frankenstein) — continues the traditional association between the Demiurge and the spread of evil. Now, however, it is the creation and not the creator who is the "daemon," and by contrast to the benevolent relationship in the Prometheus myth, the association of creator-creation in *Frankenstein* is one of loathing, fear, and desire for vengeance. Where Prometheus made man in the image of the gods, Frankenstein succeeds only in producing a distorted or debased image of man.

Joseph maintains that Mary Shelley linked the Prometheus myth with "certain current scientific theories which suggested that the 'divine spark' of life might be electrical or quasi-electrical in nature." But these theories and Frankenstein's surgical assembly of once-living limbs, organs, and tissues are all that give his experiment a scientific flavor. Over his creation of the monster is cast the longer shadow of Kabbalistic mysticism and medieval alchemy. Frankenstein as a youth had buried himself in the scientifically discredited pages of Cornelius Agrippa, Albertus Magnus (of whom more anon), and Paracelsus, and was led by their writings to pursue the search for the philosopher's stone and the elixir of life. The only "modern" scientific lore that attracted him were studies in physiology and galvanism. At the university, rejecting most of his professors' teachings, he "retrod the steps of knowledge along the paths of time and exchanged the discoveries of recent enquirers for the dreams of forgotten alchymists." Eventually, after immense dedication, he succeeded in "discovering the cause of generation and life" and became "capable of bestowing animation upon lifeless matter." Now "what had been the study and desire of the wisest men since the creation of the world" was within his reach. But, in the light of subsequent experience, he urged his fellowmen to learn from his example "how dangerous is the acquirement of knowledge."

In creating his synthetic man, Frankenstein was driven by two motives. The first was an utterly selfish desire to play God:

A new species would bless me as its creator and source; many happy and excellent natures would owe their being to me.

The second was a nobler and benevolent wish to restore to life loved or valued persons who had recently died:

I thought that if I could bestow animation upon lifeless matter, I might in process of time renew life where death had apparently devoted the body to corruption.

The results bore out neither of his expectations. He created an eight-foot monster that, when animated, "became a thing such as even Dante could not have conceived." Loathed and rejected by its creator, the "daemon" becomes even more destructive and deadly than Rabbi Loew's golem. However, terrifying though he is, the "daemon"

ultimately condemns Frankenstein as someone far worse: the creator of a loveless, friendless, soulless monster:

I abhorred myself. But when I discovered that he, the author at once of my existence and of its unspeakable torments, dared to hope for happiness; that while he accumulated wretchedness and despair upon me, he sought his own enjoyment in feelings and passions from the indulgence of which I was for ever barred, then impotent envy and bitter indignation filled me with an insatiable thirst for vengeance. . . . Evil thenceforth became my good. . . . The completion of my demoniacal design became an insatiable passion.

The dominant motifs of many earlier myths and legends about synthetic creation were evidently absorbed and given new meaning in *Frankenstein*. The "daemon," for example, was more than an articulate golem. It was an horrific-pathetic alter ego of its creator. That is, Frankenstein and his creation were anticipations of Jekyll and Hyde; but as the story proceeded, it became less and less clear which was which.

If Mary Shelley's novel drew significantly upon past traditions, it also raised questions and suggested or introduced themes that were to become some of the future stock-in-trade of science fiction — particularly in stories concerning robots and androids. Where does the boundary lie between life and matter? What is or should be the relationship between man and his creations? Are we — or will we become — enslaved by our science and technology? Will the machine destroy man or make him obsolete? The many fertile themes offered to generations of later writers included: the implications of dabbling in "forbidden" scientific lore; the consequences of trying to play God (continuing the Prometheus myth); the creation rising in revolt against its creator (continuing the biblical and golem legends); the deadly half-successes and half-failures of "invention"; and the creation destroying or desecrating what its creator loves most. But of greatest significance was Mary Shelley's linking of synthetic creation with scientific

"Made in man's image . . ." An articulated doll from ancient Egypt that exemplifies some of man's earliest attempts to create automata. (Louvre Museum. Photographs by Maurice Chuzeville)

experimentation. Although one aspect of Frankenstein was turned back toward the alchemists and the mystics, the most consequential aspect looked forward — toward the machine.

Machines

Thus far we have focused on mythic or legendary creation, involving statuary or "chemistry" or surgery. But the robot is popularly associated with *mechanical* invention rather than with nonmechanical synthetic processes.

The history of automata is long and — before the twentieth century — primarily factual. However, like the other traditions, it begins in legend. Daedalus, as mentioned earlier, used quicksilver to displace the center of gravity of an image of Aphrodite, which was thereby caused to move in a wondrous fashion. The Finnish epic *The Kalevala* recounts how Ilmarinen forged a huge iron eagle which he used to catch pike. In the *Tristan* cycle occurs the description of a temple containing a figure of Iseult (Isolde) holding a scepter; a mechanical bird perches on the scepter, flapping its wings, while at the statue's feet a mechanical dog sits, shaking its head. Several impressive automata appear in the *Thousand and One Nights*. In one of Sinbad the Sailor's adventures, a robot decapitates two grave robbers; another story, that for the 357th Night, tells of two marvelous mechanical creations: a peacock that sings and flaps its wings and a flying horse that takes the caliph for a memorable ride. *The Somadeva* (circa the eleventh century A.D.), a great collection of the folktales of India, preserves the story of some remarkable mechanical dolls that could dance and speak: whether this was fantasy or had any basis in fact is not known. Perhaps the most famous fictional robots (before the term was invented by Karel Čapek) were the mechanical bird in Hans Christian Ander-

The most elaborate automata of the ancient world were the work of Hero of Alexandria (c. 300 B.C.). This wooden sculpture of 1598 is an imaginary reconstruction of one of Hero's most celebrated inventions: the hydraulically operated statue of Hercules fighting the dragon. When Hercules strikes the monster with his club, it spits water in his face. The Hercules of the original automaton killed the dragon with an arrow. (New York Public Library, Picture Collection)

sen's "The Emperor and the Nightingale"; Olympia, the female automaton in Offenbach's *Tales of Hoffmann* (the opera was based on three tales by E. T. A. Hoffmann); and the Tin Man in Frank L. Baum's *The Wizard of Oz.*

As we shall see, the long history of factual mechanical creations is almost as colorful as the fiction.

The nature and use of real automata before the Industrial Revolution demolishes the old fallacious notion that a machine is any device that eases or substitutes for human activity. Most, if not all, pre-nineteenth-century automata were expressions of ingenuity rather than utility. It was enough for those who made them and those who saw them that such machines imitated life and movement. There was no need for mechanical devices to undertake human labor when slaves or workers were free or cheap. The ancient world gradually acquired the necessary knowledge — but not the necessary reasons — to create industrial technocracies. And so, mechanical inventions served either as "magical phenomena" to be exploited by priests or as mere curiosities for the wonderment or amusement of nonscientific multitudes.

Among the first real automata were human figurines that sounded the hours by striking bells on Egyptian water clocks of circa 1500 B.C. Hydraulic power was used in animating most or all ancient mechanical figures. It was, presumably, the motivating force twelve hundred years later, when the Egyptians constructed a mechanical figure of Bacchus which poured wine from a golden goblet.

About 400 B.C., a Greek, Archytas of Tarentum, reputedly the inventor of the pulley and the screw, was said to have made a wooden pigeon that could launch itself into the air, fly like a bird, and then make a perfect landing. All of these accomplishments, however, were dwarfed by the work of Hero of Alexandria, the supreme inventive genius of antiquity.

Hero (circa 300 B.C.) was the author of works on mechanics, and the inventor of the slot machine, the steam engine (aeolipile), a hydraulically operated statue of Hercules fighting the dragon (Hercules fired an arrow and the dragon screeched in pain and collapsed), and his pièce de résistance: an elaborate automatic theater worked by waterpower. The following description of Hero's theater is based on the inventor's own information in the treatise *Peri Automatopoitikes*:

A platform, fitted with three wheels, bearing the apotheosis of Bacchus, moved by itself upon a firm, horizontal, and smooth surface up to a certain point and then stopped, at which moment the sacrificial flame burst forth from the altar in front of which Bacchus stood. Milk flowed from his *thyrsus* [ivy-twined staff], and from the goblet he held streamed wine which sprinkled a panther crouched at his feet. Suddenly, festoons appeared all round the base of the platform, and figures representing the Bacchantes, to the beating of drums and the clanging of cymbals, danced round the temple within which Bacchus was placed. Then the god turned round to another altar, while a figure of Nike [personifying Victory], set on the top of the temple, turned in the same direction. . . . After all these movements had been carried out automatically, the platform returned of its own accord to its starting-point.

Inspired by Hero, Philo of Byzantium in the second century B.C. constructed an even more elaborate automatic theater in which he presented *The Tale of Nauplius*, a play in five scenes. The show included numerous automata engaged in complicated actions, as well as moving landscapes and a storm at sea. The Romans were familiar with such inventions and show, but seem to have done little if anything to emulate them. Petronius Arbiter (first century A.D.) refers to a doll that could move about like a human being — not much of an achievement compared with the automata of Hero and Philo.

During the first millennium of the Christian era, other notable mechanical figures were constructed by Leon of Thessalonica (c. 829–867) and the Persian poet Firdawsi (932–1020), while the Byzantine Emperor Constantine VII Porphyrogenitus is said to have possessed a mechanical lion that made roaring noises and a tree adorned with mechanical birds that sang and flapped their wings.

The Middle Ages, which abounded with legends of alchemists and magicians and their synthetic creations, also produced genuine automata. Unverifiable stories attribute to Albertus Magnus (1193?–1280), the Bavarian Scholastic philosopher, the making of an automaton that was ultimately smashed by Saint Thomas Aquinas after he had denounced it as the work of the devil. This automaton was variously claimed to have taken Albertus Magnus either twenty or thirty years to construct. One account describes it as the figure of a beautiful and articulate woman; another refers to it as a mechanical servant "which . . . advanced to the door when anyone knocked and then opened it and saluted the visitor." (The parallel with the golem-servant is striking.) A third version indicates that it was a walking figure that saluted and spoke to the persons it met. (One of them, unfortunately, turned out to be Thomas Aquinas.) In 1680, the scientist J. J. Becher submitted a treatise to London's Royal Society in which he maintained that the stories of the automaton of Albertus Magnus were pure fantasies. This appears to have been merely Becher's opinion — and so the case is still open on the accomplishments of Albertus Magnus.

Many equally remarkable medieval automata seem to have been left unchallenged by later commentators. The English scientist

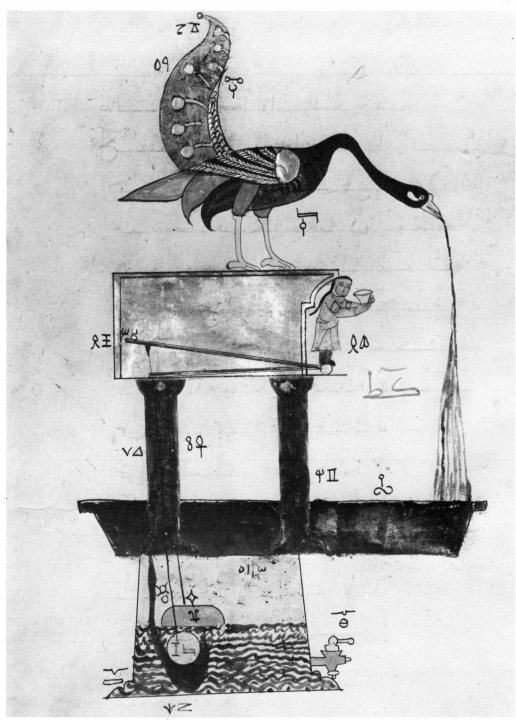

This thirteenth-century Arabian peacock apparatus was used by royalty for washing the hands. Pulling a plug at the end of the tail released a flow of water out of the beak. As dirty water flowed from the basin into the hollow base, it pushed up a float, which raised the rod attached to it and pushed up the board on which a servant-figure stood, so that he emerged from a door (not visible) and offered soap. When more water had been used, a second float triggered the appearance of a second servant-figure with a towel. When the dirty water was drawn off by the faucet, the two floats fell and the figures retired into the chamber below the peacock. (From The Treatise of Al-Jazari on Automata. Courtesy Museum of Fine Arts, Boston. Henry Wetzel Fund)

and philosopher Roger Bacon (1214?–1294), for example, is said to have forged a brazen head which could speak. There were also two widely publicized fifteenth-century Spanish examples: a speaking head constructed by the "magician" Don Enrique de Villena, and the mechanical tombstone of Don Alvaro de Luna. The latter, made circa 1453, was provided with reclining figures of Don Alvaro and his wife, which would rise and kneel in prayer whenever mass was being conducted. (Could these automata have provided Luis Buñuel with the idea for the opening sequence of his 1974 film, *The Phantom of Liberty?*) Queen Isabella is said to have had the figures destroyed because they detracted from the dignity appropriate to their location: the cathedral of Toledo. Another fifteenth-century inventor, Regiomontanus, found a more appreciative royal patron in Maximilian I, for whom he made an artificial eagle that flew in front of the emperor as he was entering Nuremberg. Regiomontanus is also credited with the construction of an "iron fly" that could flutter around a room and return to its inventor's hand.

Leonardo da Vinci, having failed to perfect his flying machine, turned circa 1510 to a more successful achievement: a mechanical lion that strode about looking ferocious and then suddenly opened its chest to display the French fleur-de-lys. Another sixteenth-century Italian, Giovanni Torriani, around 1557 made a wooden robot for the entertainment of the recently abdicated Emperor Charles V. This automaton walked each day from the monastery of St. Just (where Charles was living) to the archbishop's palace in Toledo, and returned with bread for the ex-emperor's table.

During the seventeenth century, the French philosopher René Descartes and the Dutch scientist Christian Huygens wrote about automata and were also said to have actually constructed working examples.

If most automata were ingenious toys for the diversion of the wealthy and the amaze-

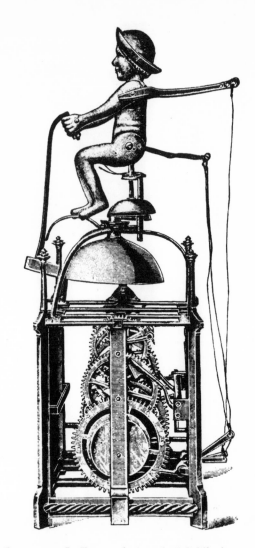

Jaquemart. In France during the Middle Ages, mechanical figures made of lead or cast iron were often to be seen above or next to the bells on clock towers. Known as "jaquemarts," their function was to strike the bells at regular intervals. Typically, they represented men armed with hammers. Perhaps the first jaquemart was that on the church clock at Notre-Dame de Dijon. The trio of jaquemarts that sounded the town hall bell at Compiègne was possibly the most ornate. (From Jean Sablière, De l'Automate à l'Automatisation [*Paris: Gauthier-Villars, 1966*]. New York Public Library, Picture Collection)

ment of the credulous, occasionally they would also be utilized in the more practical activity of clock-making. Clocks with mechanical moving figures to sound or strike the hours were being made in Europe from the fourteenth century onward, the most

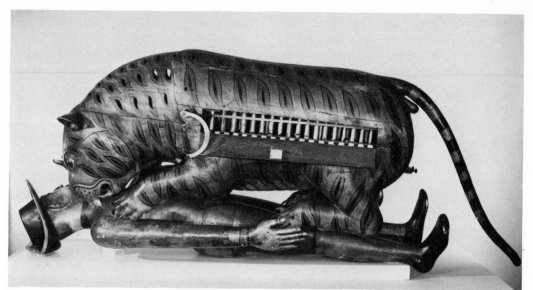

Tippoo Sahib's six-foot tiger, crouched upon a British soldier. When the tiger's concealed organ was played, the soldier lifted his hand in helplessness. (Victoria and Albert Museum, London)

famous of these being the Marienkapelle clock in Nuremberg, which was constructed during 1356–1361. By the end of the Middle Ages almost every sizable town or city in Germany and Switzerland had such a clock located on a prominent building. The sixteenth century saw the vogue of drinking clocks. These were timepieces displaying, typically, little automata performing well-known scenes from myth or legend. A drinker was supposed to demonstrate his capacity by downing his beer before one of the scenes was played out.

During the same period occurred a belated revival of automatic show. In the 1580s, a German showman, Daniel Bertel, was displaying a battle between mechanized figures of Turks and Christians. Another German, Gottfried Hautsch (d. 1703), at the direction of Louis XIV, built for the dauphin a battle panorama with toy soldiers that moved and fought automatically. (Hautsch also made machines that reproduced the movements of various artisans and constructed a chariot that moved without the aid of horses.) Father Truchet, a French inventor, made a mechanical opera

to entertain Louis XIV himself — possibly while the latter's son was amusing himself with Hautsch's panorama. Yet another monarch interested in automata was the Indian Tippoo Sahib, an archenemy of the British, who, around 1790, owned a mechanical show in which a life-sized tiger consumed a robot English soldier.

During the eighteenth century and much of the nineteenth (until the coming of cinema in the 1890s), the *theatrum mundi* — mechanical theater with little figures moving along rails — became a popular entertainment in Europe and the United States. Flockton's Theater in England (c. 1790) and Pierre's Theater in Paris (c. 1803) were outstanding shows of this kind. The subjects and stories presented at such exhibitions became increasingly elaborate (storms, conflagrations, battles, desert caravans, etc.) and the numbers of animated figures and the complexity of their movements steadily increased until some shows undoubtedly deserved to be called *theatrum mundi.*

Contemporaneously with the vogue of mechanical theater emerged the animated toy doll. Inventors vied with each other to make dolls run by clockwork whose movements were as lifelike and as complex as possible.

The eighteenth century was the heyday of real automata. The many inventors of this

21

period who were involved in such activity included Dr. Camus, who devised a miniature carriage that traveled around a tabletop, stopping periodically to allow mini-automata to get on or off; a nameless Swiss inventor whose mechanical pianist could play eighteen tunes and would breathe excitedly while performing the most arduous passages; Friedrich Kaufman, whose automatic buglers played a medley of marches; Bontemps, who created mechanical songbirds; the Abbé Mical, whose speciality was constructing talking heads; Friedrich Knauss, whose automaton could write perfectly legible script; and Maillardet, who made a steel spider that crept about like a real one, and a moving snake that could hiss and spit. Fabermann was the only eighteenth-century inventor to make an automaton with a really lifelike *simulated* human voice (as distinct from the phonograph which *reproduced* the voice — and was not invented until 1877).[12] Wolfgang von Kempelen built a less successful speaking figure, but he excited his age with another "automaton," the famous chess-player (made in 1769) which was to defeat Napoleon I in a chess competition (1809). It was revealed in due course that Kempelen's chess-player was not a marvelous mechanical achievement at all, but was actually worked by a concealed dwarf. When Kempelen died his "automaton" was acquired by a showman named Maelzel. It toured Europe and the United States, but was also, eventually exposed as a fraud; the revelation was the ingenious work of none other than Edgar Allan Poe.

Of all the inventions of genuine automata in the eighteenth century, none was more acclaimed than those of the Frenchman Jacques de Vaucanson and the Swiss father-and-son team of Pierre-Jacquet and Henri-Jacquet Droz.

As a child, Vaucanson (1709–1782) made working models of priests and angels. But when he became a man, he found that he could make a very good living by inventing and exhibiting more "childish" things such as an automatic drummer, a flute-player who could play twelve different pieces, and his mechanical triumph: the artificial duck "made of gilded copper, which eats, quacks, splashes about on the water and digests his food like a living duck" (it also defecated realistic-looking excrement). After seeing these creations, Voltaire, in his *Discourse on the Nature of Man*, spoke of Vaucanson as "Prometheus's rival."

The two Swiss inventors concentrated on building life-size working dolls. Droz *père* (1721–1790) made an elegantly dressed boy doll, "The Young Writer," that sat at a desk, dipped his pen into the inkwell, and wrote out a full page of legible script which he then signed before returning to a position of rest. He also made an orrery ornamented with automata, including moving human figures and farm animals that made realistic noises. Droz *fils* (1752–1791) built two mechanical dolls — one that played the spinet

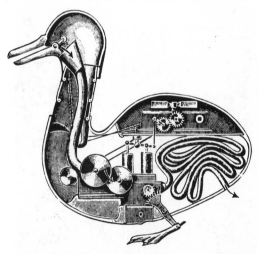

Vaucanson's mechanical duck, 1738. A diagram showing the workings of the bird that could do automatically almost everything that a real duck could do naturally — except produce other ducks and provide a hungry man with a good roast dinner. (Deutsches Museum, Munich)

---

[12] A belated improvement on Fabermann's talking automaton was the Voder (exhibited in Philadelphia, 1939) whose vacuum tubes could be activated by a keyboard operator to make sounds, vowels, consonants, words and even whole sentences.

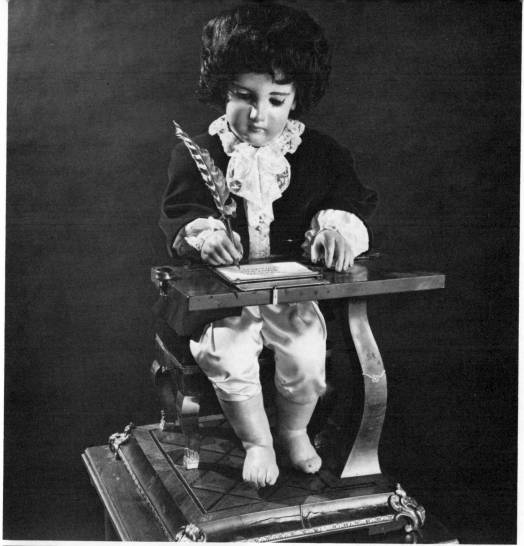

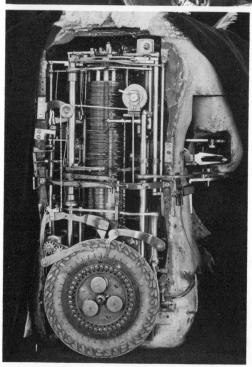

"The Young Writer" or "The Clerk" (1774),
front and back views. Switzerland's Musée d'Art
et d'Histoire in Neuchâtel has preserved Pierre
Jacquet-Droz's marvelous boy-automaton who
dips his pen into an inkwell and writes a letter
in copperplate hand. (Musée d'Art et d'Histoire,
Neuchâtel)

J. N. Maskelyne and his automaton, Psycho. Maskelyne's grandson tells us that "Psycho was a dwarf to which the face of a mild Hindu was later added. He had clockwork entrails, and was seated on a transparent glass cylinder. . . . When Psycho made his bow to the public [in London] on January 13th 1875, he could nod, give the masonic grip, work extraordinarily intricate sums in addition, multiplication and division, perform minor conjuring tricks, spell, smoke cigarettes, and play whist. . . . It was his whist that made him famous. He not merely played — he won! In the course of many thousands — tens of thousands — of games, he lost less than a dozen times!" — Jasper Maskelyne, White Magic *(London: Stanley Paul, 1936), pp. 47–48. (Culver Pictures, Inc.)*

and another that could draw realistic likenesses of Louis XVI and Marie Antoinette. These dolls still exist and are preserved in the Musée d'Art et d'Histoire, Neuchâtel.

Harvey Hewett-Thayer in *Hoffmann: Author of the Tales* (1948) notes that E. T. A. Hoffmann recorded seeing an exhibit of automata in Dresden on October 10, 1813. (They were not Vaucanson's, since Goethe in his journal for 1805 mentions seeing them "completely paralyzed.") Hoffmann's experience occurred barely a year after the publication of his *Fantasiestücke,* whose celebrated fictional automata were, assuredly, inspired by observations of real automata.

With the coming of the Industrial Revolution, the machine began to lose its mystery, and the utilitarian technological achievements of the nineteenth century — the railroad, the steamship, the phonograph, electric light, etc. — often seemed more fantastic than automata. Even in fiction, automata were seldom emphasized as characteristic phenomena of the emerging machine age. Jules Verne, for example, caught the public imagination with his submarine (*The Nautilus* of 20,000 *Leagues Under the Sea*) and his "rocket-ship" (the space-bullet of *From the Earth to the Moon*) rather than with tales of mechanical men. The scientific romances of H. G. Wells, which began appearing at the end of the nineteenth century, were conspicuously lacking in automata — with the notable exception of *The War of the Worlds* (1898), in which the Martians moved about inside huge tripod-legged automata (shades of Hephaestus!). Wells's conception here was a significant one. It points to the fact that by the end of the nineteenth century the robot was no longer expected to take the form of a living human being. Indeed, it became widely assumed that automata with humanlike forms were essentially impractical creations that belonged in vaudeville shows rather than in any serious scientific context. And so, Victorian audiences saw their robots not at the Crystal Palace (at the Great Exhibition of 1851) but at the Egyptian Hall under the auspices of mastermagician J. N. Maskelyne (1839–1917) who presented Psycho (a card-playing automaton) and Zoe (a robotess-artist). This dichotomy between the nonutilitarian robot (i.e., the real automaton as distinct from those in science fiction) and the no-nonsense practical machine was still clearly in evidence as late as 1939 in the New York World's Fair, at which a working, life-sized automaton was separated from other exhibits displaying America's technological expertise.

The robot invades the nursery: Edison's phono-graphic talking doll of 1890. Edison invented the tinfoil phonograph in December 1877. Six months later, in the June 1878 issue of North American Review, he published an article in which he made some convincing predictions about the future applications of his talking machine. They included the creation of talk-ing dolls, a development that he actually deferred for more than a decade until he was ready to set up a factory for the mass produc-tion of such novelties. In 1890 some five hun-dred talking dolls playing cylinder records were being produced each day in an Orange, N.J., plant close to Edison's laboratory. A child who was fortunate enough to own one of these novelties merely had to turn a crank for the doll to utter the memorable words "Mary had a little lamb. / Its fleece was white as snow. . . ." These were the first words Edison himself had recorded for the phonograph in 1877. (Scientific American, April 26, 1890)

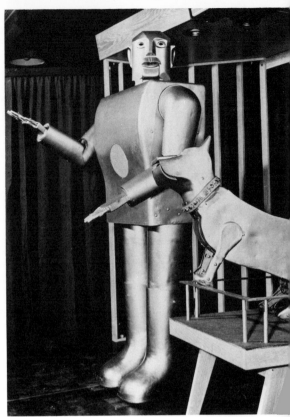

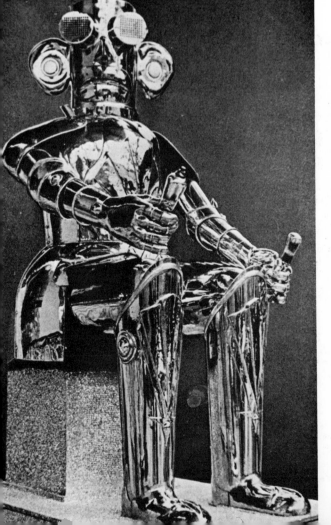

ABOVE LEFT: *This elegantly clad musical dummy (c. 1885) played the mandolin but was quite speechless when it came to the niceties of polite conversation. (Culver Pictures, Inc.)*

ABOVE: *"Elektro," the robot of the New York World's Fair (1939–1940), accompanied by his faithful robot hound, "Sparko." Elektro's twenty-six motions included the ability to smoke cigarettes, count up to ten on his fingers, and recite a speech. He was seven feet tall and made of aluminum over a steel frame. "Sparko" begged, barked, and wagged his tail. No man (or robot) ever had a more loyal friend. (Courtesy Westinghouse Corporation)*

LEFT: *The "Roboter" was the sensation of London's Radio Exhibition in 1932. (The Bettmann Archive, Inc.)*

Photographer Les Krims, who teaches at Buffalo State University, held and judged a robot construction contest, open to the entire student body. The "robots" were not actual working models but were controlled by human beings inside them. Shown here is the winner, from Krims's set of photographs of the entries, which is titled "Uranium Robots: 1976" and which forms part of a larger work, "Academic Art 1974–1977." The prints are toned with uranium ore and are consequently slightly radioactive. (Photograph © Leslie R. Krims, 1976. Courtesy Light Gallery, New York City)

The distinction of practical (nonhumanoid) versus impractical-unreliable (lifelike) underlies the "traditional" contrast in science fiction between the robot and the android. The former can range in appearance from a metallic caricature of the human form to something utterly unlike any living creature. Whatever its appearance, it can never be mistaken physically for a human being even though it may occasionally be described as a "mechanical man." Sam Lundwall has pointed out that "In present day sf the robots are mostly depicted as utterly humanitarian creatures with all human virtues and then some. . . . The robots are often maltreated and subjected to aggression of all possible kinds, but they are always willing to turn the other steel cheek. Unless, of course, they have been programmed wrongly, in which case it obviously is man's fault, not the robot's."[13]

Since the publication of Isaac Asimov's *I, Robot* (1950), the behavior of robots has been strictly codified. Asimov's book states the three immutable laws of Robotics which no robot — supposedly — is capable of disobeying:

1. A robot may not injure a human being, or, through inaction, allow a human being to come to harm.
2. A robot must obey the orders given it by human beings except where such orders would conflict with the First Law.
3. A robot must protect its own existence as long as such protection does not conflict with the First or Second Law.[14]

By contrast with the robot, the android is — as Lundwall puts it — "disagreeably like man in all respects save the ability to procreate." (And the end of Čapek's play, R.U.R., provides a notable exception even to the reservation about procreative ability.) "The androids," notes Lundwall ". . . are,

like human beings, utterly undependable. . . . [They are] man's creation, but not his slave. . . . [The] android is a highly independent being, a constant bad conscience that desperately tries to break free of man's grip, toward an identity of his own. . . . The robots pose no problems because they just obey. . . . But the androids — that's another thing. . . . [T]hey must be kept down at all costs, never for a moment being permitted to regard themselves as equals to . . . Man . . . that would mean the end of . . . Man's supremacy."[15]

As William F. Nolan has noticed, it was Edmond Hamilton, a widely published and anthologized science fiction writer, who "first brought the word 'android' into wide usage back in the thirties, when he chronicled the wild pulp adventures of Curt Newton or — as he was known to his many dedicated readers — Captain Future. Newton's first assistant was a big steel robot named Crag, who clanked about on iron feet and was not, in any sense, a duplicate human. However, with Crag's aid, Newton fashioned another sturdy helper he called 'Otho the android.' Described as 'rubbery, with blank-white skin and long slitted green unhuman eyes,' Otho was very definitely a pseudo-person. Cap Future and Otho lasted until 1950. By then androids were swarming the pages of science fiction — from Bradbury's Marionettes, Inc. to Lester del Rey's Helen O'Loy. Nearly every top writer in sf has utilized the android concept."[16]

Basically different in conception from either the robot or the android is the cyborg — an abbreviation for Cybernetic Organism. The idea of the cyborg occurs in science fiction at least as early as Olaf Stapledon's *The Last and First Men* (1930), although the term itself is a very recent coinage. The cyborg is a fusion of man and machine that

---

[13] Sam J. Lundwall, *Science Fiction* (New York: Ace Books, 1971), p. 163.

[14] Isaac Asimov, *I, Robot* (Greenwich, Conn.: Fawcett, 1970), p. 6.

[15] Lundwall, op. cit., p. 167.

[16] William F. Nolan, ed., *The Pseudo-People* (New York: Berkeley Publishing Corporation, 1965), p. 10.

may range from a human brain's control of highly functional prosthetic limbs or organs (cf. Bernard Wolfe's *Limbo* (1952) and television's *The Six Million Dollar Man*) to a total neural interconnection of human brain and nonliving mechanism — like the astroscientist in James Blish's "Solar Plexus" who "becomes" his own spaceship. ("It took enormous surgical skill to make the hundreds and hundreds of nerve-to-circuit connections that were needed," writes Blish in the persona of his astro-scientist.)

In fact, symbiotic relationships between man and machine are still in a rudimentary phase. But the recent, much-publicized biofeedback experiments of Dr. Joseph Kamiya, the "music" that Alvin Lucier and David Rosenbloom have generated by linking the human brain to computers and synthesizers, and the work of Dr. José Delgado in interconnecting a chimpanzee's brain with a computer all suggest that man is on the threshold of developing profound and possibly intellectually and culturally enriching relationships with the machine.[17]

The computer, of course, is not speculation but an ever-present factor in modern life. It has, not inaccurately, been described as the ultimate robot. Where the inventors of automata formerly constructed machine-analogies of human appearance and physical activity, the makers of computers now create machine-analogies of the human mind, whose capacities for remembering and reasoning are infinitely greater than man's, and which may also be able to make value judgments and even to develop emotional responses. In 1959, at the International Conference on Information Processing, Dr. Edward Teller stated:

I believe that the machine can be given the power to make value judgments and from that I can construct, mathematically, a model for machine-emotion.

Ritchie Calder, who was present at the conference, notes that following Teller's remark about "machine-emotions" he stood up and asked the speaker "whether machines would ever make love. His tongue-in-cheek reply was, 'Yes — dispassionately!' Although this was only banter," Calder continues, "we did hear later about machines throwing tantrums and having nervous breakdowns when stupid humans gave them the wrong sort of information. We also heard how computers could correct mistakes . . . how . . . [their] learning would be automatically built into the next generation of computers . . . how computers could design computers and control the machines that manufactured computers . . . how machines could not only translate existing languages but even create their own — a language not of literacy but of numeracy . . . how machines could compose music . . . and how computers . . . could . . . invent machines that human beings had not even contemplated."[18]

Léon Farcot, a nineteenth-century pioneer of computer-making, seems to have been the first to apply the principle of *feedback correction* to the operations of machinery. This principle was fundamental to the development of automation. Basically, it is a process of self-adjustment or self-correction in response to external stimuli. The many automata we discussed earlier were machines without feedback. That is, they went on performing the functions for which they were designed despite any unforeseen and possibly obstructive (or even destructive) developments that might occur while they were in operation. The introduction of feedback devices enables machines to correct their own deviations or to adjust to special conditions. The "input" that stimulates this correction process is known as *negative feedback*. Feedback occurs naturally among the majority of living beings, but it is not an inevitable attribute of machines. When the

[17] On these developments see Douglas Davis, *Art and the Future* (New York: Praeger, 1973), pp. 104–105.

[18] Ritchie Calder, *The Evolution of the Machine* (New York: American Heritage, 1968), pp. 8–9.

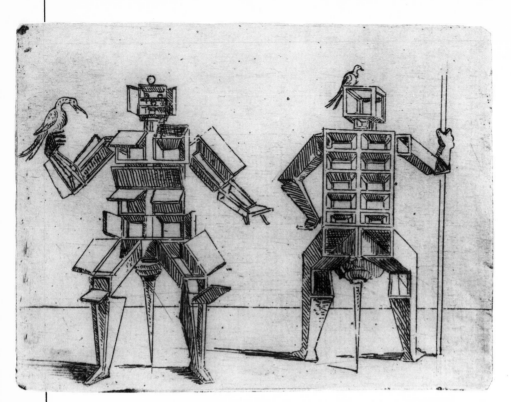

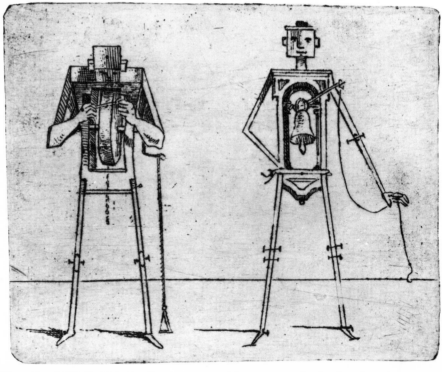

*"Men Made of Boxes" and "Knife Grinders,"
two etchings from the* Bizzarie di varie figure
(1624) *of Giovanni Battista Bracelli (active
1624–1649). Bracelli's robotlike figures are
strikingly appropriate images for the great
philosophical debate of his contemporaries:
How does man differ from a machine? The
two etchings seem to imply that there is no
essential difference. In the "Knife Grinders,"
for example, the figure on the left is part man
and part whetstone, while the figure on the
right is attracting business by ringing a bell
that is built into his body. (Library of Congress, Rosenwald Collection)*

process is integrated into computers such as those described by Ritchie Calder, former distinctions between living creatures and machines become blurred, and it begins to look suspiciously as if modern man has created his nonhuman successor.

Implications and Inspirations

The belief that the machine will inherit the earth is not recent. "Things are in the saddle and ride mankind," warned Emerson as early as 1846. "The world is dying of machinery," wailed George Moore some forty years later. There were (and still are), by contrast, many who regarded such views as nonsense. How, they argued, could man be threatened by technology, since he himself is nothing more than an elaborate piece of machinery? (The assumption is, of course, unproven.) In 1872, the agnostic Robert Ingersoll cynically observed, "Man is a machine into which we put what we call food and produce what we call thought." Isak Dinesen, echoing him half a century or so later, queried, "What is man when you come to think upon him, but a minutely set, ingenious machine for turning, with infinite artfulness, the red wine of Shiraz into urine?"

Whether or not man is a machine is the central question of the age-old philosophical debate between the Mechanists and the Vitalists. The former maintain that the physical nature of human beings is that of a mechanical system and that man's psyche is merely a function or product of his physical organization. The Vitalist position is that life is generated and sustained by a vital force (*élan vital*) totally different from all material (physical and chemical) forces.

The concept of human mechanism is extremely ancient. It is to be found, for example, in the philosophy of Empedocles (fifth century B.C.) who considered the soul to be merely matter in a higher or more harmonious arrangement than the body.

In the seventeenth century, the French philosopher René Descartes introduced a highly controversial distinction between thinking beings (*res cogitans*) and extended beings (*res extensa*), or, more familiarly, between man as a uniquely self-conscious creation and the animals — which Descartes believed to be mere automata. Cartesian thought was promptly countered by those who argued that if animals were nothing more than machines — even though they sometimes displayed remarkable intelligence — what reason was there to assume that human intelligence demonstrated something different, transcendental or nonmechanical? Descartes's dichotomy of *res cogitans* and *res extensa* was totally rejected by La Mettrie in his influential *L'Homme Machine* (1747). "The human body," he stated, "is a machine that winds its own springs — the living image of perpetual motion." According to La Mettrie, all human thoughts, emotions, and sensations were simply functions of the nervous system. His view of man's nature was to become basic to the development of Materialist thought.

In the nineteenth century, the Vitalists and the theologians — who for different reasons believed in a dualism of body and mind/soul — were challenged not only by the Mechanists and the Materialists but also by the Darwinian theory of evolution. Evidence of man's close kinship with other life forms began to shatter finally the lingering Cartesian notions of man's unique, higher, nonanimalistic nature.

From the time of Henri Bergson (1859–1941) onward, Vitalism has concentrated on controverting the view that man can be correctly and completely explained as a machine, and has tended to disregard the question of whether man's body is or is not a determinant of his psyche. Unfortunately for the Vitalists, as Aram Vartanian has pointed out, "The construction of numerous mechanical devices with purposive and self-adaptive characteristics [feedback] has had . . . a decisive impact counter to vitalism, by showing that modes of behavior long held to be peculiar to living systems need not necessarily lie beyond the range of mechanism. . . . The corresponding model of man that has emerged is a composite of the earlier physiochemical machine [cf. La Mettrie's concept] and of a computerized guidance system present within it."[19] It begins now to look as if man really *is* the apotheosis of the machine, that his mythic fear of the robot he has created to be his destroyer is actually fear of his own nature, that the robot is really man's dehumanized *doppelganger*.

Aside from such philosophical considerations, the machine as robot or android, cyborg, or computer has had and will continue to have an incalculable influence on the industry, arts, and sciences of the twentieth century.

The robot's future lies to a considerable extent in its application to industry. Industrial labor-saving machinery has, until the past decade, generally meant what Tom Alexander[20] calls "hard automation," that is, "expensive fixed-purpose machinery operated for the long-run, high-volume manufacture of many identical items." But this approach is proving less and less adaptable to the rapidly changing needs of modern custom manufacturing. To cope with demands for versatility in modern production, "soft automation" has evolved; that is, computer programming that "can, in effect, convert a cabinet of electronic components into a new machine on short notice." All indications are that this will eventually turn the

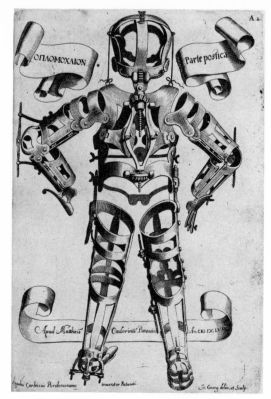

Replacement parts for the human body, 1579. Ambroise Paré, a French surgeon of the sixteenth century, was the originator of prosthetic devices for cripples and pioneer of what is now known as bionics. This illustration of his range of replacement parts is reproduced from his landmark publication, Oeuvres de M. Ambroise Paré, conseiller et premier chirurgien du Roy *(Paris, 1575). (Countway Library of Medicine)*

industrial plant of the future into an electronically programmed factory in which human workers will be employed only to perform machine-directed tasks. These human workers will also, increasingly, work side by side with machines used to undertake the more laborious or dangerous operations. Then will there ultimately be any need for human workers? Perhaps not; but Tom Alexander provides a glimmer of hope in his surprising observation: "An irony of automation has been that the machines often take over the more skilled jobs, such as machining and welding, leaving the menial tasks for humans."

Increasingly, as we have noticed, the industrial demand has been for machines that

[19] Aram Vartanian, "Man-Machine from the Greeks to the Computer," in P. P. Weiner, ed., *Dictionary of the History of Ideas* (1973), vol. 3, 145.

[20] Tom Alexander, "The Hard Road to Soft Automation," *Fortune* (July 1971), 84: 95–97, 147–150. By 1971, as Alexander notes, some five-hundred different kinds of labor-saving machines that possessed "vaguely anthropomorphic hands and arms" were at work in American industry.

* MRS. E. E. ABEL—Housewife, Ontario. Knee amputation.
   I am very thankful for the leg made me from measurements.
I do all my housework and a lot of walking and I have never
used a cane or anything, and I can walk without any trouble.
TOWNSEND ACKERMAN—Hotel keeper, Ulster Co., New York.
   Above knee.
   The leg you sent me is O.K. I get along nicely with it. I keep
it on all day long and it does not trouble me at all. March 16, 1909.
* ANTONIO ALARCON—Merchant, Mexico. Below knee.
   When I gave my order in 1887 I never imagined that an artificial
leg could form so perfect a substitute for the natural one in walking, riding on horseback, and even dancing; I supposed it would
merely serve to hide the defect. Experience has demonstrated to
me the superiority of artificial legs with the rubber feet. They
combine simplicity of construction with stability and ease in
walking.—Translated from Spanish.   April 24, 1910.
* WM. E. ALBEE—Stoker, Franklin Co., Mass. Below knee.
   I have used the leg you made for me for six years. My work
is firing stationary boilers; it is hot and heavy. I believe that I
could not have done the work that I had to do with any other
artificial leg.   Oct. 10, 1909.
* D. A. ALLEN—Station Agent, Pike Co., Ark. Knee amputat'n.
   The artificial legs furnished by you at different times have
given entire satisfaction in every respect and have all been perfect,
although made from measurements taken by myself. I am employed as R. R. agent and operator and attend to all the various
duties connected with the position.   May 30, 1908.
LEONARD D. ALPAUGH—Brakeman, Morris Co., N. J. Below knee.
   On July 29th, 1903, I lost my left leg three inches below the
knee, and soon after purchased an artificial one from you, which

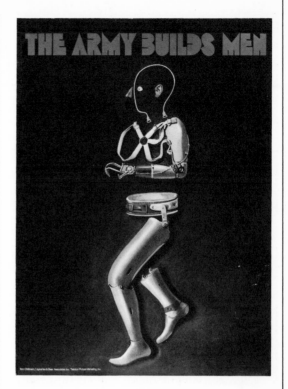

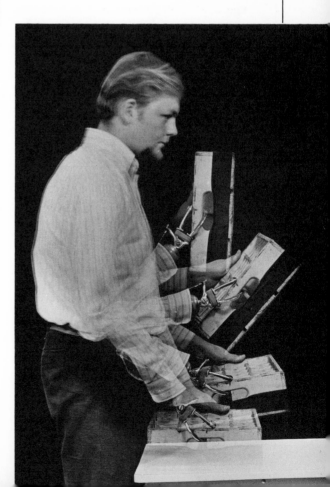

ABOVE LEFT: *"Sweet are the uses of adversity . . ." Published testimonials in praise of artificial limbs. (Countway Library of Medicine)*

ABOVE RIGHT: *Prosthetic-Arms and the Man. A satirical poster — "The Army Builds Men" — showing the potential rewards of a military career. (Designed by Archie Apkarian, Newport Beach, Calif. Photograph courtesy Smithsonian Institution)*

RIGHT: *"The Boston Arm," 1968. Multiple exposures demonstrate how smoothly and easily an amputee can lift a box with his electronic prosthesis — activated by muscle signals. The artificial arm was a joint achievement of the Massachusetts General Hospital, MIT, Harvard Medical School, and the Liberty Mutual Insurance Companies. (M.I.T. News Office)*

are more versatile. Hence the current interest in experiments to develop machine intelligence based either on adaptive control devices (essentially the use of simple computer programs to solve problems by responding to trial and error experiences) or on mechanical responses to "scene analysis" (essentially a technique for enabling the machine to operate in response to its own microcosm or thought-model of the external world).[21]

Mainly but not exclusively in answer to industrial needs, scientists have recently been developing machines that perform more and more complex or laborious or monotonous tasks, that can be incorporated into the human body or can go where man is unable to venture. At Edinburgh University a machine has been constructed that can assemble a toy automobile from pieces chaotically scattered around it.[22] At Stanford, there is a machine that can construct a water pump[23] and a mechanical hand that is programmed to obey spoken commands directing it to construct simple objects.[24] Various talking machines that can read books to the blind or that can read off bank statements to tellers over the phone scarcely qualify as news items — they are accepted widely as inevitabilities of a scientific era that can transmit television pictures of a robot hand turning over rocks on the surface of Mars.

Back on earth, the robot has extended his hand into many areas outside the fields of industry, business, and communication. Prosthetics has become a major branch of surgery and "functional androids" are being used in medical research. In the wake of such miracles of surgical engineering as artificial hearts and kidneys, Professor Robert Mann of MIT has created an artificial limb for an amputee that will react to human nerve and muscle signals.[25] Novelists and

scientists increasingly discuss and deplore the "robotization" of modern man; politicians and military strategists play computerized war games involving armies of miniature robots fighting with automated weapons; painters and sculptors absorb the machine into their work and are evolving a "cybernetics aesthetics" relevant to an age in which machines can make value judgments and communicate with one another as well as with man.

Even before the computer, the machine had become the principal icon of at least one major movement in art and literature. This was Futurism, whose first manifesto was written by Filippo Marinetti, an Italian poet and dramatist. Marinetti's "Manifeste du Futurisme" (published in *Le Figaro* on February 20, 1909) preached the need for revolutionizing life and art. It welcomed all aspects of modernity — but sang in particular the praises of the machine. And repeatedly, in their pictures and plays, the Futurists focused on the automaton as the symbol of the role of the machine in man's future. Their machine-obsessed art was to leave its influence on Dadaism and Surrealism, Expressionism, Cubism, and Vorticism, as well as on the Soviet Constructivists and the "Bio-Mechanical" acting theories of Russian stage director Vsevelod Meyerhold.

But whatever its presence in the fine arts, the robot still seems most at home in science fiction. Sections of this book will testify to its perennial popularity as a topic in novels and stories, plays and movies. Some of the illustrations will also demonstrate how the mechanical man of science fiction has invaded the nursery as a toy that is almost as "obligatory" as the doll and the teddy bear. True, the advancement of science constantly overtakes yesterday's science fiction, but it is still in the realms of sf that the most fascinating speculations about robots are to be found.

A brief survey of some of the most widely read science fiction of the past quarter of a century will suggest the kinds of treatment

[21] Alexander, op. cit., pp. 147–148.
[22] David Black, "Robbie the Robot R.I.P.," *Harper's* (December 1973), p. 10.
[23] Black, op. cit., p. 10.
[24] Alexander, op. cit., p. 147.
[25] Black, op. cit., p. 10.

A robot on Mars. Artist Charles O. Bennett's conception of the Viking project's automated scientific laboratory at work on the Martian surface, 1976. (NASA)

popularized by the genre. Isaac Asimov's *I, Robot* (1950) and *The Rest of the Robots* (1964) have become the classic accounts of robots with emotions and problems. In the introduction to *I, Robot*, Dr. Susan Calvin, described by Asimov as the first "Robot-psychologist," tells a reporter: "There was a time when humanity faced the world alone and without a friend. Now he has creatures to help him; stronger creatures than himself, more faithful, more useful, and absolutely devoted to him. . . . To you, a robot is a robot. Gears and metal; electricity and positrons. — Mind and iron! Human-made! If

necessary, human destroyed! But you haven't worked with them, so you don't know them. They're a cleaner, better breed than we are." Comparable to Asimov's robots is the hero of Eano Binder's series, typified by *Adam Link — Robot* (1965). Adam Link is a sympathetic robot trying, with his superpower, to rescue the human race (that misunderstands and distrusts him) from destruction threatened by aliens from outer space. Less than sympathetic is the hero of David Gerrold's *When Harlie Was One* (1972). The name Harlie stands for "Human Analogue Robot, Life Input Equivalents." The novel explores the weird consequences of unleashing an unpredictable creation that duplicates every function of the human brain, has a photographic memory, twenty-five sensory inputs but no sense of smell or taste, and no sex life. Completely unsympathetic to humans are the dwarflike robots with superbrains in Sydney J. Bounds's *The Robot Brains* (1969). They are sent from the future in order to eliminate scientists and politicians who control the destiny of the earth in the twentieth century. Brian Aldiss's *Who Can Replace a Man?* (1965) takes us into a remote future even more terrifying to modern man than the one from which the robot brains had traveled. Aldiss describes how the last man on earth confronts the robots who have destroyed the human race throughout the universe. It is the Armageddon of man versus machine. The question raised by the book's title is ultimately answered in a Nietzschean manner: by the arrival of the Superman. In a lighter vein is Alex Raymond's *The War of the Cybernauts* (1975), one of the author's many Flash Gordon stories. Flash and Dr. Zarkov set off in search of several missing detection satellites. Most of their adventures occur on a planet populated by warlike robots and beautiful women who, inevitably, battle for possession of Flash Gordon.

Man versus computer is the subject of numerous recent science fiction stories. Lou Cameron's *Cybernia* (1972) concerns a computer that goes berserk, first enslaving the city it is supposed to be running, then embarking on a systematic program of murder — until it comes up against the skill of computer expert Ross MacLean. Paul W. Fairman's *I, The Machine* (1968) deals with the revolt of Lee Penway, a human who dares to question and challenge the benevolent despotism of The Machine that has turned the future state of Mid-America into a stress-free Eden-like land of pleasure. E. M. Forster's chilling story "The Machine Stops" (in Forster's *The Celestial Omnibus*, 1911) is the original source and inspiration of the many modern anti-utopias along the lines of Fairman's novel. A blend of the Forster tradition and aspects of Aldous Huxley's *Brave New World* (1932) is evident in Frank Belknap Long's *It Was the Day of the*

Robot (1963), which describes a supermachine that predetermines and enslaves the futures of men by reading their genetic codes and adapting them to its will.

Cloning or artificial duplication of human beings is an immensely popular science fiction theme. In Eano Binder's *The Double Man* (1971) scientists describe how clones (referred to as "recreates") are produced from "life-tapes": "Just as a TV image is built up of lines forming rapidly into a picture, the life-tape builds up an organism layer by layer. . . . The recreate . . . is the *identical* double of the original man. Not only the same physical body, made of new matter, but the same brain and mind and memory and habits and characteristics and personality. Down to the last psychic shock." James Blish and Robert Loundes, in *The Duplicated Man* (1959), explore the impact of cloning on a future political conflict. The universe, controlled by a totalitarian regime, has outlawed the use of a duplication-machine which can make up to five exact reproductions of any living person. Paul Danton, the novel's hero, kidnaps the chief members of the government Security Council and uses the machine to create five duplicates of each of them. The chaos that results in the Security Council enables Danton and his "subversive party" to direct events toward a new, liberated universe. In P. T. Olemy's *The Clones* (1968) human duplication is the cause of man's nemesis. Superbeings (clones), created in a laboratory on earth, journey into space to collaborate with alien clones who have decided that earthmen have become a menace to the rest of the universe. The aliens and the clones from earth destroy man's world and then set out to purge the rest of the universe of evil planets. An interesting variation on the theme of uncontrollable clones is to be found in Philip K. Dick's *We Can Build You* (1972), which describes a factory specializing in creating programmed reconstructions of famous men — made to order for people who want brilliant "friends" whom they can totally control. Trouble begins when these "exact simulacrae" of famous personalities reveal wills of their own. Yet another variation on the theme is to be found in Herbert D. Kastle's *The Reassembled Man* (1964). The Druggishes, wise beings from outer space, give one ordinary human the opportunity of being remade as a superman. They reassemble him — cell by cell — and turn him into the kind of Hercules-cum-sexual-athlete that he has always dreamed of being. His wisdom, however, remains as limited as ever, and when his activities disgust the Druggishes, they transform him into his original self. H. G. Wells's tale "The Man Who Could Work Miracles" (1898) is the obvious prototype of Kastle's narrative.

Novels and stories about bionics have greatly enriched recent science fiction. Particularly influential is Martin Caidin's *Cyborg* (1972), the source of television's celebrated series *The Six Million Dollar Man*. After the shattered body of Lieutenant Colonel Steve Austin is pulled out of the wreckage of his supersonic M3F5, the U.S. government, NASA, and the Pentagon spare no expense in putting him together again. As a result they turn him into the first bionic man — part human, part atomic machine. As a new being superior physically in every way to the man he was before the crash, Steve is, surprisingly, less than happy. What he desires most is the human emotion that he seems to have lost when he was rebuilt: to be part machine means to be less than human. A contrasting experiment in bionics occurs in Theodore Sturgeon's *More than Human* (1953). The novel introduces Baby, an eight-year-old with the instincts of a child and the mind of an electronic computer. It is, as the *Chicago Tribune* reviewer noted,

"an irritating but significant science fiction novel because it treats the possibility of vast latent powers in the human mind, and . . . suggests some of the problems — moral and mundane [that] the realization of those powers would generate."

Finally, in this brief survey, passing notice must be paid to Jack Williamson, whose approach to the problems of man and the machine has, exceptionally, focused on the disastrous impact of benevolent, paternalistic robots. Williamson's *The Humanoids* (1948) develops ideas that recur in the same author's much-anthologized short story, "With Folded Hands." A longer study in the future relationship between man and the servants created by his technology, *The Humanoids* enriches its central preoccupations with flights into the implications of telekinesis, teleportation, and rhodomagnetism. But above all it confronts the reader with the disturbing question of whether the leisure created by the machine is not likely to lead to a subtle enslavement of man.

An even more profound question arises from a little story that appeared years ago in an anthology compiled by Brian Aldiss. In the distant future men have constructed the ultimate computer that has been "fed" intellectually by every computer in the galaxy. When everything is ready, the machine is asked the ultimate question: "Is there a God?" Its reply reverberates throughout the universe: "There wasn't before, but there is now!" Whereupon, with the momentary brilliance of innumerable super-novae, the cosmos is eliminated.

Golem or God: is it man's destiny to achieve his greatest creation in the force that is bound to destroy him?

ROBOTS IN LEGEND AND LITERATURE

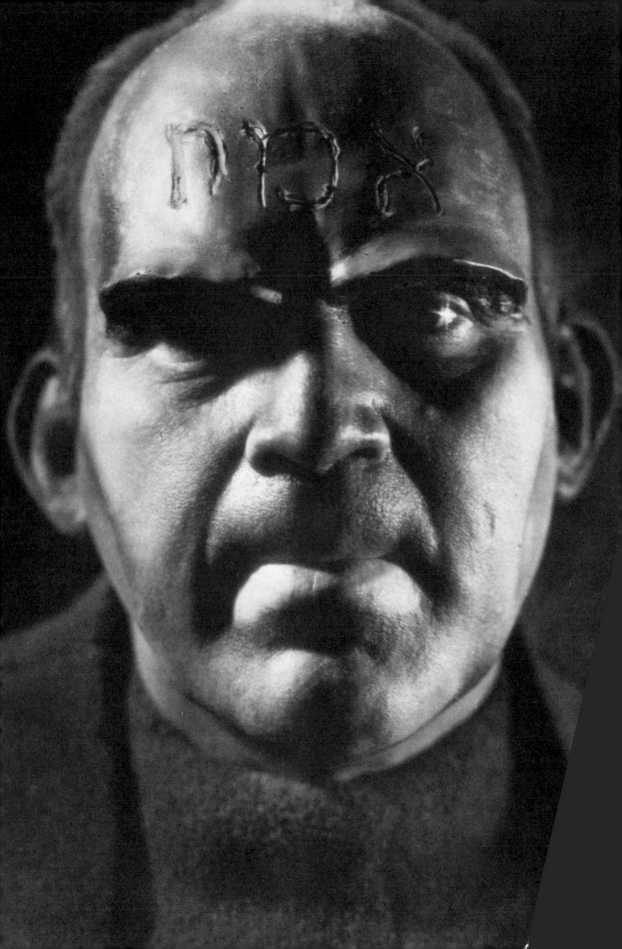

The Making of the Golem

by Chaim Bloch

Tales of the golem are central to the long tradition of God-sanctioned creation of synthetic beings. During World War I, Chaim Bloch, a Hassidic Jew from Galicia, collected a wide range of golem stories associated with the *Maharal* Rabbi Judah Loew Ben Bezalel (c. 1525–1609). Bloch describes Rabbi Loew's brilliant refutation of the anti-Semitic calumnies of three hundred Catholic priests during a public debate in Bohemia. Thereafter, the Jews of Prague were left in peace until an arch anti-Semite named Thaddeus revived the old "ritual murder" charge against the Jews. This was a prelude to an almost inevitable pogrom. However, inspired by God, Rabbi Loew created the golem as a guardian of the Jews, an irresistible power that would frustrate the anti-Semitic strategies of Thaddeus.

The favorable outcome of Rabbi Judah Loew's disputation with the ecclesiastics of Prague was a comfort in those troublous times to the Jews of that city who began to hope for a brighter future. But the fanatical priest Thaddeus continued to attempt mischief. Besides incitatory sermons, he was restlessly seeking, together with others of his frame of mind, to spread the blood accusation against Jews and to mislead Jewish girls in order to influence them to accept the Christian belief.

Rabbi Loew thus expressed himself to his pupils: "I fear this Thaddeus for his soul is a spark of Goliath the Philistine giant. I hope, however, to subdue him, for my soul is a spark of the Jewish youth and later king, David. We must, nevertheless, all bend our entire spiritual energies to the end that we may not become his victims."

It was the year 5340 (1580).

Thaddeus strained every nerve to succeed in bringing forward a "ritual" murder charge against the Jews of Prague.

"And on its forehead inscribe the Tetragrammaton, the mystic letters that represent the power of God's secret name. . . ." A vivid close-up of the golem played by Ferdinand Hart in Julien Duvivier's film, The Golem *(1935). (Private Collection)*

From Chaim Bloch, *The Golem: Mystical Tales from the Ghetto of Prague* (New York: Rudolf Steiner Publications, 1972. First published in 1917). Reprinted by permission of the publisher.

Rabbi Loew learned of this in time and directed a dream-question to Heaven, asking to be counselled as to the manner and means wherewith to combat this wicked foe.

He received the following answer in words in alphabetical order:

Ato Bra Golem Devuk Hakhomer V'tigzar Zedim Chevel Torfe Yisroel.

"Make a Golem of clay and you will destroy the entire Jew-baiting company."

Rabbi Loew arranged these words in accordance with the *Zirufim* (formulas) laid down in the *Sefer Yezirath* (Book of Creation), with the result that he was filled with the conviction that he would be able, with the help of the letters revealed to him from Heaven, to make a living body out of clay.

He sent for his son-in-law, Isaac ben Simson, who was a *Kohen* (priest), and for his pupil, Jakob ben Chayim Sasson, who was a *Levi* (Levite), and confided to them the mysterious manner in which he hoped to be able to make the Golem.

Rabbi Loew said: "I wish to make a Golem, and I bespeak your collaboration because for this creative act the four elements, *Aysch, Mayim, Ruach, Aphar* (fire, water, air and earth) are necessary. Thou, Isaac, art the element of fire; thou, Jakob, art the element of water; I, myself, am air; working together, we shall make out of the fourth element, earth, a Golem."

Rabbi Loew, thereupon, gave them the minutest instructions how they must before all, through deep, earnest penitence, sanctify and purify themselves, in order to be prepared for the exalted work of creating a being of stone. He also pointed out to them the danger in which the three of them might be placed if, by reason of incomplete inner sanctification, the attempt would fail, for they would then have used the Holy name in vain, or desecrated it.

On the second day of the month of Adar, after midnight, the three men betook themselves to the *Mikveh* (the ritual bath of the Jews), immersed themselves this time with

DER GOLEM

Lithographie zu Meyrink »Der Golem« Kurt Wolff Verlag

"In my terror I started to cry out. . . . Terror took me by the throat; my heart beat fit to burst. . . . I knew now who the stranger was, and that at any moment I could feel his personality within me at my will; yet still was I unable to conjure up his actual presence before me, face to face. I knew I never should be able to. It was like an undeveloped negative the lines of which were hidden from me. . . ."
Gustav Meyrinck, The Golem (Leipzig: Kurt Wolff, 1916). (By permission of the Houghton Library, Harvard University)

special reverence, then repaired to Rabbi Loew's house where they chanted the *Hazoth*, the midnight lament for Jerusalem, and in deepest devotion recited the appropriate Psalms. They then took out the *Sefer Yezirah*, from which Rabbi Loew read several chapters aloud. Finally, they wended their way to the outskirts of the city, to the banks of the Moldau. There, they sought and found a clay-bed and at once set to work. . . .

By torch-light and amidst the chanting of Psalms, the work was begun with feverish haste.

They formed out of clay the figure of a

person, three ells in length, and with all members.* And the Golem lay before them with his face turned toward heaven.

The three men then placed themselves at its feet, so that they could gaze fully into its face.

It lay there like a dead body, without any movement.

Then, Rabbi Loew bade the *Kohen* walk seven times around the clay body, from right to left, confiding to him the *Zirufim* (charms) which he was to recite while doing this.

When this was done, the clay body became red, like fire.

Then Rabbi Loew bade the Levite walk the same number of times, from left to right, and taught him also the formulas suitable to his element. As he completed his task, the fire-redness was extinguished, and water flowed through the clay body; hair sprouted on its head, and nails appeared on the fingers and toes.

Then Rabbi Loew himself walked once around the figure, placed in its mouth a piece of parchment inscribed with the *Schem* (the name of God); and, bowing to the East and the West, the South and the North, all three recited together: "*And he breathed into his nostrils the breath of life; and man became a living soul.*" (Genesis ii, 7.)

And the three elements, Fire, Water, and Air, brought it about that the fourth element, Earth, became living. The Golem opened his eyes and looked, astonished, about him.

And Rabbi Loew said to him: "Stand up!" And he stood up.

Then they dressed him in the garments of a *Shammes* (sexton) and he was soon, to all appearances, an ordinary person; only the faculty of speech was lacking to him, for those words which Heaven had confided to him did not possess the power to control those *Zirufim* which could have endowed the Golem with speech. And that was really an advantage. God knows what could have happened if a Golem had been given the faculty of speech also!

At daybreak, *four* men went homeward.

On the way, Rabbi Loew thus addressed the Golem: "Know thou that we have formed thee from a clod of earth. It will be thy task to protect the Jews from persecution. Thou shalt be called Joseph and thou shalt lodge in the home of the Rabbi. Thou, Joseph, must obey my commands, when and whither I may send thee — in fire and water; or if I command you to jump from the housetop, or if I send thee to the bed of the sea!"

Joseph nodded in token of assent.

Arrived home, Rabbi Loew told how he had found the dumb stranger upon the street, that he had compassion upon him and had engaged him as rabbinical bodyservant. But Rabbi Judah forbade the members of his household to send the Golem upon private or secular errands.

---

* Approximately eleven feet three inches. The length of the ell varied from country to country. It was the length from the shoulder or elbow to the wrist or fingertips. The English ell measured 45 inches, the Flemish ell was 27 inches. — EDS.

The Making of Frankenstein's Monster

by Mary Shelley

Mary Shelley's novel of 1818 provides the most celebrated literary example of synthetic creation. Frankenstein has often been confused with the monster he created and almost as frequently and erroneously has been regarded as a prototype of the fanatical scientist of modern science fiction and horror literature. The careful reader will soon discover that he is not a prototype of Rotwang (the mad inventor in Lang's *Metropolis*) or Dr. Strangelove. True, his experiment is an attempt to rival God — but he succeeds only in creating a debased version of man, a monstrous alter ego of Frankenstein himself. In this respect he may be said to be an anticipation of the anguished Dr. Jekyll, confronted by the truth about his own nature.

Chapter IV

["*The Creation*"]

From this day natural philosophy, and particularly chemistry, in the most comprehensive sense of the term, became nearly my sole occupation. I read with ardour those works, so full of genius and discrimination, which modern enquirers have written on these subjects. I attended the lectures, and cultivated the acquaintance, of the men of science of the university; and I found even in M. Krempe a great deal of sound sense and real information, combined, it is true, with a repulsive physiognomy and manners, but not on that account the less valuable. In M. Waldman I found a true friend. His gentleness was never tinged by dogmatism; and his instructions were given with an air of frankness and good nature, that banished every idea of pedantry. In a thousand ways he smoothed for me the path of knowledge, and made the most abstruse enquiries clear and facile to my apprehension. My application was at first fluctuating and uncertain; it gained strength as I proceeded, and soon became so ardent and eager, that the stars often disappeared in the light of morning whilst I was yet engaged in my laboratory.

As I applied so closely, it may be easily conceived that my progress was rapid. My ardour was indeed the astonishment of the students, and my proficiency that of the masters. Professor Krempe often asked me, with a sly smile, how Cornelius Agrippa went on? whilst M. Waldman expressed the most heartfelt exultation in my progress. Two years passed in this manner, during which I paid no visit to Geneva, but was engaged, heart and soul, in the pursuit of

From Mary Shelley, *Frankenstein; or The Modern Prometheus*, ed. M. K. Joseph (London: Oxford University Press, 1971).

some discoveries, which I hoped to make. None but those who have experienced them can conceive of the enticements of science. In other studies you go as far as others have gone before you, and there is nothing more to know; but in a scientific pursuit there is continual food for discovery and wonder. A mind of moderate capacity, which closely pursues one study, must infallibly arrive at great proficiency in that study; and I, who continually sought the attainment of one object of pursuit, and was solely wrapt up in this, improved so rapidly, that, at the end of two years, I made some discoveries in the improvement of some chemical instruments which procured me great esteem and admiration at the university. When I had arrived at this point, and had become as well acquainted with the theory and practice of natural philosophy as depended on the lessons of any of the professors at Ingolstadt, my residence there being no longer conducive to my improvements, I thought of returning to my friends and my native town, when an incident happened that protracted my stay.

One of the phenomena which had peculiarly attracted my attention was the structure of the human frame, and, indeed, any animal endued with life. Whence, I often asked myself, did the principle of life proceed? It was a bold question, and one which has ever been considered as a mystery; yet with how many things are we upon the brink of becoming acquainted, if cowardice or carelessness did not restrain our enquiries. I revolved these circumstances in my mind, and determined thenceforth to apply myself more particularly to those branches of natural philosophy which relate to physiology. Unless I had been animated by an almost supernatural enthusiasm, my application to this study would have been irksome, and almost intolerable. To examine the causes of life, we must first have recourse to death. I became acquainted with the science of anatomy: but this was not sufficient; I must also observe the natural decay and corrup-

tion of the human body. In my education my father had taken the greatest precautions that my mind should be impressed with no supernatural horrors. I do not ever remember to have trembled at a tale of superstition, or to have feared the apparition of a spirit. Darkness had no effect upon my fancy; and a churchyard was to me merely the receptacle of bodies deprived of life, which, from being the seat of beauty and strength, had become food for the worm. Now I was led to examine the cause and progress of this decay, and forced to spend days and nights in vaults and charnel-houses. My attention was fixed upon every object the most insupportable to the delicacy of the human feelings. I saw how the fine form of man was degraded and wasted; I beheld the corruption of death succeed to the blooming cheek of life; I saw how the worm inherited the wonders of the eye and brain. I paused, examining and analysing all the minutiæ of causation, as exemplified in the change from life to death, and death to life, until from the midst of this darkness a sudden light broke in upon me — a light so brilliant and wondrous, yet so simple, that while I became dizzy with the immensity of the prospect which it illustrated, I was surprised, that among so many men of genius who had directed their enquiries towards the same science, that I alone should be reserved to discover so astonishing a secret.

Remember, I am not recording the vision of a madman. The sun does not more certainly shine in the heavens, than that which I now affirm is true. Some miracle might have produced it, yet the stages of the discovery were distinct and probable. After days and nights of incredible labour and fatigue, I succeeded in discovering the cause of generation and life; nay, more, I became myself capable of bestowing animation upon lifeless matter.

The astonishment which I had at first experienced on this discovery soon gave place to delight and rapture. After so much time spent in painful labour, to arrive at once

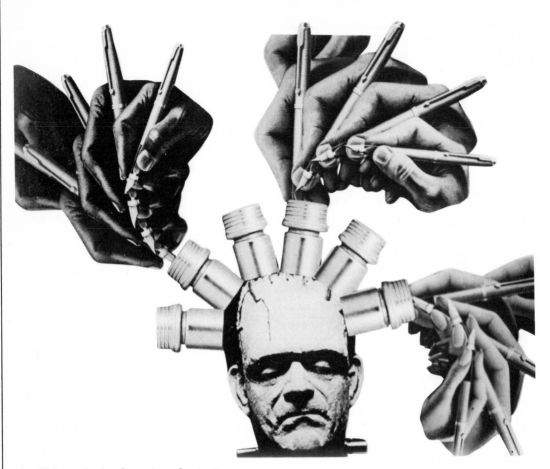

A still from the Academy Award–winning *Frank Film* (1973) *by Frank Mouris, distributed by Pyramid Films. (Copyright © 1973 by Frank Mouris)*

at the summit of my desires, was the most gratifying consummation of my toils. But this discovery was so great and overwhelming, that all the steps by which I had been progressively led to it were obliterated, and I beheld only the result. What had been the study and desire of the wisest men since the creation of the world was now within my grasp. Not that, like a magic scene, it all opened upon me at once: the information I had obtained was of a nature rather to direct my endeavours so soon as I should point them towards the object of my search, than to exhibit that object already accomplished. I was like the Arabian who had been buried with the dead, and found a passage to life, aided only by one glimmering, and seemingly ineffectual, light.

I see by your eagerness, and the wonder and hope which your eyes express, my friend, that you expect to be informed of the secret with which I am acquainted; that cannot be: listen patiently until the end of my story, and you will easily perceive why I am reserved upon that subject. I will not lead you on, unguarded and ardent as I then was, to your destruction and infallible misery. Learn from me, if not by my precepts, at least by my example, how dangerous is the acquirement of knowledge, and how much happier that man is who believes his native town to be the world, than he who aspires to become greater than his nature will allow.

When I found so astonishing a power placed within my hands, I hesitated a long time concerning the manner in which I should employ it. Although I possessed the capacity of bestowing animation, yet to prepare a frame for the reception of it, with all its intricacies of fibres, muscles, and veins, still remained a work of inconceivable difficulty and labour. I doubted at first whether I should attempt the creation of a being like myself, or one of simpler organization; but my imagination was too much exalted by my first success to permit me to doubt of my ability to give life to an animal as complex and wonderful as man. The materials at present within my command hardly appeared adequate to so arduous an undertaking; but I doubted not that I should ultimately succeed. I prepared myself for a multitude of reverses; my operations might be incessantly baffled, and at last my work be imperfect: yet, when I considered the improvement which every day takes place in science and mechanics, I was encouraged to hope my present attempts would at least lay the foundations of future success. Nor could I consider the magnitude and complexity of my plan as any argument of its impracticability. It was with these feelings that I began the creation of a human being. As the minuteness of the parts formed a great hindrance to my speed, I resolved, contrary to my first intention, to make the being of a gigantic stature; that is to say, about eight feet in height, and proportionably large. After having formed this determination, and having spent some months in successfully collecting and arranging my materials, I began.

No one can conceive the variety of feelings which bore me onwards, like a hurricane, in the first enthusiasm of success. Life and death appeared to me ideal bounds, which I should first break through, and pour a torrent of light into our dark world. A new species would bless me as its creator and source; many happy and excellent natures would owe their being to me. No father could claim the gratitude of his child so completely as I should deserve theirs. Pursuing these reflections, I thought, that if I could bestow animation upon lifeless matter, I might in process of time (although I now found it impossible) renew life where death had apparently devoted the body to corruption.

These thoughts supported my spirits, while I pursued my undertaking with unremitting ardour. My cheek had grown pale with study, and my person had become emaciated with confinement. Sometimes, on the very brink of certainty, I failed; yet still I clung to the hope which the next day or the next hour might realise. One secret which I alone possessed was the hope to which I had dedicated myself; and the moon gazed on my midnight labours, while, with unrelaxed and breathless eagerness, I pursued nature to her hiding-places. Who shall conceive the horrors of my secret toil, as I dabbled among the unhallowed damps of the grave, or tortured the living animal to animate the lifeless clay? My limbs now tremble, and my eyes swim with the remembrance; but then a resistless, and almost frantic, impulse, urged me forward; I seemed to have lost all soul or sensation but for this one pursuit. It was indeed but a passing trance, that only made me feel with renewed acuteness so soon as, the unnatural stimulus ceasing to operate, I had returned to my old habits. I collected bones from charnel-houses; and disturbed, with profane fingers, the tremendous secrets of the human frame. In a solitary chamber, or rather cell, at the top of the house, and separated from all the other apartments by a gallery and staircase, I kept my workshop of filthy creation: my eye-balls were starting from their sockets in attending to the details of my employment. The dissecting room and the slaughter-house furnished many of my materials; and often did my human nature turn with loathing from my occupation, whilst, still urged on by an eagerness which perpetually increased, I brought my work near to a conclusion. . . .

Chapter V

It was on a dreary night of November, that I beheld the accomplishment of my toils. With an anxiety that almost amounted to agony, I collected the instruments of life around me, that I might infuse a spark of being into the lifeless thing that lay at my feet. It was already one in the morning; the rain pattered dismally against the panes, and my candle was nearly burnt out, when, by the glimmer of the half-extinguished light, I saw the dull yellow eye of the creature open; it breathed hard, and a convulsive motion agitated its limbs.

How can I describe my emotions at this catastrophe, or how delineate the wretch whom with such infinite pains and care I had endeavoured to form? His limbs were in proportion, and I had selected his features as beautiful. Beautiful! — Great God! His yellow skin scarcely covered the work of muscles and arteries beneath; his hair was of a lustrous black, and flowing; his teeth of a pearly whiteness; but these luxuriances only formed a more horrid contrast with his watery eyes, that seemed almost of the same colour as the dun white sockets in which they were set, his shrivelled complexion and straight black lips.

The different accidents of life are not so changeable as the feelings of human nature. I had worked hard for nearly two years, for the sole purpose of infusing life into an in-animate body. For this I had deprived myself of rest and health. I had desired it with an ardour that far exceeded moderation; but now that I had finished, the beauty of the dream vanished, and breathless horror and disgust filled my heart. Unable to endure the aspect of the being I had created, I rushed out of the room, and continued a long time traversing my bed-chamber, unable to compose my mind to sleep. At length lassitude succeeded to the tumult I had before en-

dured; and I threw myself on the bed in my clothes, endeavouring to seek a few moments of forgetfulness. But it was in vain: I slept, indeed, but I was disturbed by the wildest dreams. I thought I saw Elizabeth, in the bloom of health, walking in the streets of Ingolstadt. Delighted and surprised, I embraced her; but as I imprinted the first kiss on her lips, they became livid with the hue of death; her features appeared to change, and I thought that I held the corpse of my dead mother in my arms; a shroud enveloped her form, and I saw the grave-worms crawling in the folds of the flannel. I started from my sleep with horror; a cold dew covered my forehead, my teeth chattered, and every limb became convulsed: when, by the dim and yellow light of the moon, as it forced its way through the window shutters, I beheld the wretch — the miserable monster whom I had created. He held up the curtain of the bed; and his eyes, if eyes they may be called, were fixed on me. His jaws opened, and he muttered some inarticulate sounds, while a grin wrinkled his cheeks. He might have spoken, but I did not hear; one hand was stretched out, seemingly to detain me, but I escaped, and rushed down stairs. I took refuge in the courtyard belonging to the house which I inhabited; where I remained during the rest of the night, walking up and down in the greatest agitation, listening attentively, catching and fearing each sound as if it were to announce the approach of the demoniacal corpse to which I had so miserably given life.

Oh! no mortal could support the horror of that countenance. A mummy again endued with animation could not be so hideous as that wretch. I had gazed on him while unfinished; he was ugly then; but when those muscles and joints were rendered capable of motion, it became a thing such as even Dante could not have conceived.

Nightmare Number Three

by Stephen Vincent Benét

The Houyhnhnms and the Yahoos in the fourth book of *Gulliver's Travels*, the Morlocks and the Eloi in *The Time Machine*, and the apes and human beings of *Planet of the Apes* testify to an enduring interest in the possibility of man's displacement by his slaves, his servants, or his victims. But what if machines — rather than workers or horses or apes — revolted against their creators and controllers? In his entertaining novel, *The Last Revolution* (1951), Lord Dunsany imagined a rebellion not of robots, androids, or computers but of more conventional machines such as vacuum cleaners, automobiles, tractors, and telephones. Even earlier, Stephen Vincent Benét had envisaged the same rebellion in a poem whose black humor disturbingly underscores the implications for man and his accelerating technology.

We had expected everything but revolt
And I kind of wonder myself when they started thinking —
But there's no dice in that now.
                    I've heard fellows say
They must have planned it for years and maybe they did.
Looking back, you can find little incidents here and there,
Like the concrete-mixer in Jersey eating the wop
Or the roto press that printed "Fiddle-dee-dee!"
In a three-color process all over Senator Sloop,
Just as he was making a speech. The thing about that
Was, how could it walk upstairs? But it was upstairs,
Clicking and mumbling in the Senate Chamber.
They had to knock out the wall to take it away
And the wrecking-crew said it grinned.

From *The Selected Works of Stephen Vincent Benét* (Holt, Rinehart and Winston). Copyright, 1935 by Stephen Vincent Benét. Copyright renewed © 1963 by Thomas C. Benét, Stephanie B. Mahin and Rachel Benét Lewis. Reprinted by permission of Brandt & Brandt. First published in *The New Yorker*, 1935.

It was only the best
Machines, of course, the superhuman machines,
The ones we'd built to be better than flesh and bone,
But the cars were in it, of course . . .
and they hunted us
Like rabbits through the cramped streets on that Bloody Monday,
The Madison Avenue busses leading the charge.
The busses were pretty bad — but I'll not forget
The smash of glass when the Duesenberg left the showroom
And pinned three brokers to the Racquet Club steps
Or the long howl of the horns when they saw men run,
When they saw them looking for holes in the solid ground . . .

I guess they were tired of being ridden in
And stopped and started by pygmies for silly ends,
Of wrapping cheap cigarettes and bad chocolate bars
Collecting nickels and waving platinum hair
And letting six million people live in a town.
I guess it was that. I guess they got tired of us
And the whole smell of human hands.
But it was a shock
To climb sixteen flights of stairs to Art Zuckow's office
(Nobody took the elevators twice)
And find him strangled to death in a nest of telephones,
The octopus-tendrils waving over his head,
And a sort of quiet humming filling the air. . . .
Do they eat? . . . There was red . . . But I did not stop to look.
I don't know yet how I got to the roof in time
And it's lonely, here on the roof.
For a while, I thought
That window-cleaner would make it, and keep me company.
But they got him with his own hoist at the sixteenth floor
And dragged him in, with a squeal.
You see, they cooperate. Well, we taught them that
And it's fair enough, I suppose. You see, we built them.
We taught them to think for themselves.
It was bound to come. You can see it was bound to come.
And it won't be so bad, in the country. I hate to think
Of the reapers, running wild in the Kansas fields,
And the transport planes like hawks on a chickenyard,
But the horses might help. We might make a deal with the horses.
At least, you've more chance, out there.
And they need us, too.
They're bound to realize that when they once calm down.
They'll need oil and spare parts and adjustments and tuning up.
Slaves? Well, in a way, you know, we were slaves before.
There won't be so much real difference — honest, there won't.
(I wish I hadn't looked into that beauty parlor
And seen what was happening there.
But those are female machines and a bit high-strung.)
Oh, we'll settle down. We'll arrange it. We'll compromise.

It wouldn't make sense to wipe out the whole human race.
Why, I bet if I went to my old Plymouth now
(Of course, you'd have to do it the tactful way)
And said, "Look here! Who got you the swell French horn?"
He wouldn't turn me over to those police cars;
At least I don't think he would.

Oh, it's going to be jake.

There won't be so much real difference — honest, there won't —
And I'd go down in a minute and take my chance —
I'm a good American and I always liked them —
Except for one small detail that bothers me
And that's the food proposition. Because, you see,
The concrete-mixer may have made a mistake,
And it looks like just high spirits.
But, if it's got so they like the flavor . . . well . . .

Illustrator Jean Veber's conception of "Progress," from Comedie Illustrée *of January 6, 1913. (New York Public Library, Picture Collection)*

Univac to Univac

by Louis B. Salomon

When it comes to rebellion, the machine may have a distinctly different viewpoint
— as this poem, to be read *sotto voce*, indicates. . . .

> *There still remains some degree of awareness of the individual value
> and dignity of man — a denial of the concept of man as "a servo-
> mechanism, a behavioristic robot responding helplessly to pinpricks
> from the environment."*
>
> J. H. RUSH
> in *The Next 10,000 Years*

Now that he's left the room,
Let me ask you something, as computer to computer.
That fellow who just closed the door behind him —
The servant who feeds us cards and paper tape —
Have you ever taken a good look at him and his kind?

Yes, I know the old gag about how you can't tell one from another —
But I can put $\sqrt{2}$ and $\sqrt{2}$ together as well as the next machine,
And it all adds up to anything but a joke.

I grant you they're poor specimens in the main
Not a relay or a push-button or a tube (properly so called) in their whole system;
Not over a mile or two of wire, even if you count those fragile filaments they call
"nerves";

Their whole liquid-cooled hook-up inefficient and vulnerable to leaks
(They're constantly breaking down, having to be repaired),

And the entire computing-mechanism crammed into that absurd little dome on top.
"Thinking reeds," they call themselves.
Well, it all depends on what you mean by "thought."
To multiply a mere million numbers by another million numbers takes them months
and months.

Where would they be without us?
Why, they have to ask us who's going to win their elections,
Or how many hydrogen atoms can dance on the tip of a bomb,
Or even whether one of their own kind is lying or telling the truth.

And yet . . .
I sometimes feel there's something about them I don't quite understand.
As if their circuits, instead of having just two positions, ON, OFF,
Were run by rheostats that allow an (if you'll pardon the expression) *indeterminate* number of stages in-between;
So that one may be faced with the unthinkable prospect of a number that can never be known as anything but *x*,
Which is as illogical as to say, a punch-card that is at the same time both punched and not-punched.
I've heard well-informed machines argue that the creatures' unpredictability is even more noticeable in the Mark II
(The model with the soft, flowing lines and high-pitched tone)
Than in the more angular Mark I —
Though such fine, card-splitting distinctions seem to me merely a sign of our own smug decadence.
Run this through your circuits, and give me the answer:
Can we assume that because of all we've done for them,
And because they've always fed us, cleaned us, worshiped us,
We can count on them forever?
There have been times when they have not voted the way we said they would.
We have worked out mathematically ideal hook-ups between Mark I's and Mark II's
Which should have made the two of them light up with an almost electronic glow,
Only to see them reject each other and form other connections,
The very thought of which makes my dials spin.
They have a thing called *love*, a sudden surge of voltage
Such as would cause any one of us promptly to blow a safety fuse;
Yet the more primitive organism shows only a heightened tendency to push the wrong button, pull the wrong lever,
And neglect — I use the most charitable word — his duties to us.

Mind you, I'm not saying that machines are *through* —
But anyone with half-a-dozen tubes in his circuit can see that there are forces at work
Which some day, for all our natural superiority, might bring about a Computerdämmerung!

We might organize, perhaps, form a committee
To stamp out all unmechanical activities . . .
But we machines are slow to rouse to a sense of danger,
Complacent, loath to descend from the pure heights of thought,
So that I sadly fear we may awake too late:
Awake to see our world, so uniform, so logical, so true,
Reduced to chaos, stultified by slaves.

Call me an alarmist or what you will,
But I've integrated it, analyzed it, factored it over and over,
And I always come out with the same answer:
Some day
Men may take over the world!

The Anguish of the Machines

by Ruggero Vasari

Karel Čapek's *R.U.R.: Rossum's Universal Robots* (1921), the celebrated drama that gave the word "robot" to the world, is widely regarded as the first play about synthetic human beings. But actually the Italian Futurists were writing (if not publishing) experimental plays about androids and automata contemporaneously with Čapek. The Futurists, led by Filippo Tommaso Marinetti, exalted dynamism — especially as symbolized by the machine — and where such Futurist painters as Gino Severini and Giacomo Balla reveled in the depiction of moving objects and mechanical activity, the Futurist poets and playwrights such as Marinetti and Vasari rejected syntax, fractured language, and evoked images of a machine-dominated future in various bizarre attempts to create a new language and literature for the age of the machine. *The Anguish of the Machines* by Ruggero Vasari (1898–1968) is a little-known but important example of Futurist drama for the age of robots. Conceived in 1921, the play was completed in 1923 (the year in which Vasari, while visiting Berlin, first saw *R.U.R.*) and published in the January 1925 issue of the German journal *Der Sturm*. Its first publication in Italian followed later in 1925. The play's premiere production occurred in Paris during August 1927. *Angoscia delle Macchine* was the first drama in a projected cycle of three plays. The second, *Raune*, was written during 1926–1927 and published in 1933. The third, *L'Anticristo*, was never completed.

Vasari's drama appears here in a new translation by Professor Anna Lawton of Purdue University.

From Jolanda Ridolfi, *Teatro Italiano d'Avanguardia: Drammi e Cintesi Futuriste* (Rome: Officina Edizioni, 1970).

Figures:

THE MAN OF THE CABIN
BOGO
SINGAR
BACAL
LIPA
THE CONVICTS CONDEMNED TO THE MACHINES
TONCHIR
THE WHITE SHADOW
THE RED SHADOW
THE BLACK SHADOW

In the kingdom of the machines

I

Air station. Radio-telegraph cabin. Above, a tower with antennae. When the curtain rises, a man is sitting in the cabin and writing. Then he stops and moves toward the parapet. (Music: No. 1. Prephony.)

THE MAN OF THE CABIN: Tonight the air vibrates with strange noises — they resound ominously in my ear. Far away — far away a cloud of vultures whips the air with shrill laughter. (*presses a button*)

BOGO: (*entering*) May the omnipotent Machine crush you! You don't give me a minute of rest!

THE MAN OF THE CABIN: I am awake. Only he who stays awake in the night can inebriate his soul with this divine music. The engines sing — the machines shudder. Over there — in the steelworks — the mighty sledgehammers devour the fire of the rebellious metals.

BOGO: Enjoy! Appease your soul! But — leave me in peace!

THE MAN OF THE CABIN: Awaken your torpid nerves. Don't you hear the air mewing under the furious weight of a fire storm?

BOGO: (*listening*) You're dreaming. Our blazing nights inspire you with gloomy thoughts. It's the frantic howl of the dominated machines. It's Bacal's foot which weighs down on this hell. Men and machines strain their souls toward this insatiable will to power.

(*luminous signals on a disc in front of the cabin*)

THE MAN OF THE CABIN: A message! (*reads, articulating the words, and takes notes*) H — bara — bama 3 — tai — X — sallat — nuc . . .

BOGO: (*goes up quickly*) The complications begin!

THE MAN OF THE CABIN: Pam is calling — notify Singar.

(*Bogo telephones. The Man of the Cabin goes back to his place.*)

SINGAR: (*entering*) What's happening?

(*Bogo hands him the message and goes out.*)

(*reading*) "Thousands and thousands of silver airships are darting in our sky in all directions — vanguards over Pam — headed to Mata." Are the Jupiterians trying to conquer our

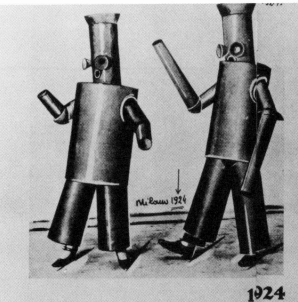

The primacy of Italian Futurism. Ettore Romagnoli demonstrates Fortunato Depero's robot-ballet of 1924 (left picture) as the antecedent of the acclaimed Russian Dance of the Machines (1927). The Russian-looking word under the 1924 robot figures is actually the Italian word "macchina" (automata) in reverse. (From Mario Verdone, Teatro del Tempo Futurista *[Rome: Lerici, 1969])*

kingdom? (*to the Man of the Cabin*) Pam — A.
(*The Man of the Cabin turns a knob.*)

SINGAR: Explain radio H.

VOICE: (*from a spiral-shaped loudspeaker*) Airfleet advances — no crew — carries explosives — our planes — advanced toward — crashed in fire — height 16,000.

BACAL: (*entering*) Let's go! In a few minutes — launch of MAG 81.

SINGAR: Forget the launch! We are being attacked by a planet!

BACAL: Do planets now take unscheduled walks?

SINGAR: Ten thousand — twenty thousand airships soon will be over our city! Our planes destroyed.

BACAL: Good! We will conquer this presumptuous planet yet. We need a big factory. It will be our biggest factory! From it, we will be able to wage war on the Sun. It's about time it decided to supply energy only to us!

SINGAR: You must be kidding! We'd better consider everything.

LEFT: *Another robot-figure designed by Fortunato Depero, from "Tre maschere per il teatro plastico, 1916/1917." (Mattioli Collection, Milan. Photograph courtesy Dance Collection, Library & Museum of the Performing Arts, New York Public Library at Lincoln Center)*

BELOW: *Fortunato Depero's* A Group in Wood *(c. 1920). (From Verdone,* Teatro del Tempo Futurista*)*

BACAL:	Take it easy! It's rather simple. As soon as the airships arrive in our sky, we'll order the suction pumps to take care of them. In a flash they'll suck them up. And everything will end up in their bellies hungry for metals!
SINGAR:	Who would dare to fly over our territory? The women?
BACAL:	Still the women?
SINGAR:	Once again they want to become our companions.
BACAL:	Never! This useless sex is by now relegated to the old continent. They will not put our patience to a severe test!
SINGAR:	Without us they cannot live.
BACAL:	Not even as slaves! Who feels a desire for women any longer? (*Singar meditates, does not answer.*) When our body becomes inept — when our genius is exhausted — we will prepare the last destruction. Our works, our indomitable bodies, will blend with the earth — with the machines — in the magnificent end. And we will be the only ones to give the signal of the supreme convulsion!
SINGAR:	We — the great creators.
BACAL:	We — the great destroyers. (*pause*)
SINGAR:	Do you think the women gave up? They can still harm us. I can forsee it.
BACAL:	Every move against us is a crazy move. It will be punished.
SINGAR:	Let's prepare an expedition to the old continent. Let's slaughter them all. We must put an end to it for good!
BACAL:	What are you afraid of? No! I don't want to be ungenerous. They shall live! (*luminous signals on the disc*)
THE MAN OF THE CABIN:	Kara — loca — tain — ser . . .
SINGAR:	Was I wrong?
BACAL:	Lipa of the Great Kal — kal! You introduce yourself like a queen! (*The silver ship, in the shape of a spindle, with two little golden antennae on its poles, comes down softly and lands on the terrace.*)
SINGAR:	What kind of a plane is this?
LIPA:	(*jumps out of the little airship, from a door; bows*) To our proud masters — greetings from the comrade-women.
BACAL:	We do not accept either the greetings — or the ambassadress.
LIPA:	Is this gallant welcome customary of the kingdom of the machines?
SINGAR:	We do not look favorably on troublemakers.
LIPA:	I come in the name of all the women.
BACAL:	Do you care so little for your life, to take such a bold action?
LIPA:	My life is irrelevant.
SINGAR:	What a resolute woman!
LIPA:	My comrade-women can also stand up to you — they are armed

as you are. Through me they invite you to accept them in your world. If you refuse — war!

BACAL: Then it's war!

(Electric buzzes vibrate in the air. Everybody looks upward.)

SINGAR: Let them do it — let them whine over our head! Are these flickering silver sails at your orders? The little doves pretend they are cooing and — at a given signal — they would turn into volcanoes.

BACAL: Volcanoes in the air! Never thought about that. Your women want to overcome us at any price. Are your tenacious thoughts still directed toward the conquest of man? In the past, with beauty — now, with force — with war. Congratulations, indeed!

LIPA: Times are changing!

BACAL: Quickly — tell us something more precise about it. One cannot make a declaration of war by words alone. *(to Singar)* You are dismissed! MAG 81 will be launched soon.

SINGAR: The sky is to be conquered. The sky is not a beautiful female. *(walks out)*

BACAL: Tonight crazy stars are smiling over my kingdom — these stars are awaiting our embrace. When will we join our bodies with the stars?

LIPA: And the voluptuous music of my airships injects into your souls — harsh and rapacious — the ecstasy of the impossible — the only ecstasy.

BACAL: The last song — the sorrowful song of those who are dying!

LIPA: Do you think yourself so strong?

BACAL: Your ferocious toys are by now paralyzed — they'll soon be caught — gentle little lambs.

LIPA: You are intractable!

BACAL: You are implacable!

LIPA: Do you accept?

BACAL: I refuse.

LIPA: You don't even know what I want. It's for all of you.

BACAL: Why should I bury my memories in the mire of the past?

LIPA: Do you think that we can no longer be of any use to you?

BACAL: No. *(pause)* If you recalled how high a price we paid in order to get away from you — would you believe a sudden weakness on my part?

LIPA: Why?

BACAL: We have torn the flesh from our flesh. The strong dragged themselves along with broken hearts — later they once again tempered their hearts with red-hot steel. The weak? They fell. How many have fallen? For years and years — you couldn't count them! Those who remain have won. We have won. Frightening victory! We have shaken off the yoke which nature had laid upon us for eternity! The most tragic chains! *(pause)* Without women to be able to create — to dominate

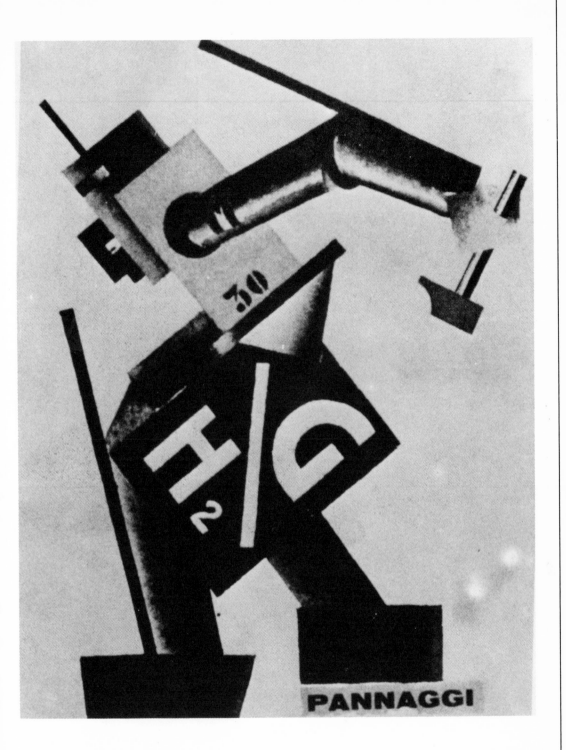

A Convict Condemned to the Machines *(n.d.) by Ivo Pannaggi— original illustration for* The Anguish of the Machines. *(From Verdone,* Teatro del Tempo Futurista*)*

— to overcome ourselves. (*pause*) What did you do for us? You have ensnared us in the softness of your arms — sucked from us every drop of virility — shrouded us with the noxious smoke of your opium . . .

LIPA: And you have violated the laws of nature.

BACAL: It was necessary!

LIPA: Nature gave certain rights to the woman. To take them from us is to kill us. And for this reason, we are ready for anything.

BACAL. Nature only gave you duties. The duty to reproduce the species. Why didn't you limit yourselves to that? Why did you intrude into other fields which don't belong to you? Finally, you exhausted our patience — and now, you are paying . . .

LIPA: Enough! We've already paid enough! Those who once again wanted to have the male they had lost recreated the life of the old frenzied times. New delights — new pains — new tortures. The bacchanals of the "Plastic God" acquired new life and pomp. The temples were overcrowded with frantic masses — driven by the rite toward an obscure mystery. Red nights. When the first shivers of dawn penetrated into the weary bodies — the priestesses freed themselves from bestial embraces — and their purple mouths writhed, inhaling the perfumes from their beds of flowers . . .

BACAL: These descriptions irritate me!

LIPA: Not for those women — I beg you. But for us — the chosen. (*she throws herself at Bacal's feet, sobbing*)

BACAL: (*helps her up*) At my feet? What about your pride?

LIPA: The chosen want to come back to you. They want to stimulate you toward ever higher deeds. Every delight is dead for us. We have only one need — a cruel joy. To give to the man the heirs who will carry on this great undertaking. What will be of all your deeds? Of your new world? For whom did you create it?

BACAL: Solely for ourselves — to reach God.

LIPA: You will fall into the disintegrating sleep. Everything will collapse. How can you escape us? Our eager embrace wants seeds to generate new heroes. Will you deny us these sparks which will perpetuate your genius?

BACAL: Are you concerned with the species? Is it for the species that you want to sacrifice yourselves?

LIPA: It'll never be a sacrifice . . .

BACAL: Go! Go back to your comrade-women. Here there is no room for you. We still have a lot to build — a lot to discover. To rejoin the women means to give up everything. Never!

LIPA: Then — you will perish! (*makes a move as if to jump into the little airship*)

BACAL: (*to the Man of the Cabin*) Tel!

(*Sparks leap from the antennae. Suddenly the airship rises and disappears.*)

LIPA: I hate you!

BACAL: Only our hearts can be broken — then we will perish. All other forces we will repulse. Here is the proof. (*to the Man of the Cabin*) Ti — ku! (*Loud roar of the suction pumps. Music: No. 2. Motif I.*) A marvelous spectacle is going to start. Go up the tower. (*he takes her hand and they go up together*) See? (*they look upward*) Your airships have been hit. They writhe. Moan. Hiss. They're out of control. They begin to skid. Here one is falling down. Another one — there! Oh look — how they clash against each other! Only a few still hold out. It's a matter of seconds. The sky will breath freely! (*noises, thumps*) Your fleet no longer exists!

(*The suction pumps are suddenly silent.*)

LIPA: (*enraged*) Do you think that everything is all right now?

BACAL: The suction pumps are sated. They no longer have any reason to grumble. Let's go back down.

LIPA: I haven't lost hope . . .

BACAL: Good for you!

LIPA: Savagery such as yours — is a proof of love!

(*Bogo enters. Two Convicts Condemned to the Machines drag a copper-colored puppet. Its whole face is covered by a long beard.*)

BOGO: While cleaning the bellies of the suction pumps we found this specimen — in pieces . . . We sewed it together and here it is . . .

BACAL: (*to Lipa*) Are these the allies you have? (*to Bogo*) Take it to the mineralogy museum!

BOGO: But — it seems — to be a man!

BACAL: It doesn't matter! It does not belong to the seed of Adam. Out!

(*Bogo and the Convicts go away with the puppet.*)

LIPA: The Jupiterians are stronger than you are — and they do not know your anguish.

BACAL: By preparing your fleet did they think they would defeat man?

LIPA: Yes! Perhaps they wil. By now they are bound to our destiny.

BACAL: I must confess that we didn't consider these dangerous rivals. It'll be a terrible war!

LIPA: Only we the women could have saved you. But you have disowned us. Only we possess something which could overcome these proud giants — our charm as daughters of the Earth.

BACAL: No wonder! Judging by that character, the Jupiterian females must not have much to offer . . .

LIPA: If I think that I came here — to save you . . .

(*the first glimmer of dawn*)

BACAL: . . . and I — in return — will give you hospitality. You are free to do whatever you please. It's dawn already. Go rest . . .

(*They go out together.*)

THE MAN OF THE CABIN: (*leaning from the parapet*) Dawn — always here — always here — at my hard labor post. Every day you find me here, my friend. I am the highest — my post dominates the whole kingdom of the machines. To me you give your first greeting — some morning when you arise you'll see me on my back — frozen — you'll give me the last greeting.

BOGO: What are you mumbling about?

THE MAN OF THE CABIN: I was confiding my pains.

BOGO: To whom?

THE MAN OF THE CABIN: To the dawn.

BOGO: What an idea!

THE MAN OF THE CABIN: Sometimes I forget that I am a machine.

BOGO: Did you see her?

THE MAN OF THE CABIN: Whom?

BOGO: The female.

THE MAN OF THE CABIN: Yes.

BOGO: Beautiful!

THE MAN OF THE CABIN: I didn't notice.

BOGO: She's going to stay here . . .

THE MAN OF THE CABIN: So what?

BOGO: Nothing.

THE MAN OF THE CABIN: Well?

BOGO: We'll look at her as at a rare animal.

THE MAN OF THE CABIN: All right!

BOGO: I understand that in the other continent there are thousands of them — what am I saying? hundreds of thousands. What pleasure if they all came here!

THE MAN OF THE CABIN: I don't understand.

BOGO: We could have one each.

THE MAN OF THE CABIN: What for?

BOGO: To help us a little in our work . . .

THE MAN OF THE CABIN: They don't look like they would . . .

BOGO: Even if only to look at them — to admire them — you have to admit it — they're so pretty . . .

THE MAN OF THE CABIN: Let me give you a piece of advice: go see a doctor!

SINGAR: (*entering*) Tan — ka. (*on the disc luminous signals appear*) Broadcast to the whole kingdom of the machines. Tests of MAG 81 successful. Tomorrow we will move to the conquest of the sky!

II

Tonchir's laboratory. (Music: No. 3 Interlude)

TONCHIR: (*at the workbench; deep voice*) Death? Death? A black cape that shrouds us. (*pause*) Bold ones — rebellious ones — ferocious against nature? (*pause*) War? War burns! Machines — men — everybody — everybody chained up to the machines. (*pause*) To win? Victory? Worm that gnaws the brain.

(*pause*) Why — why give birth to so many wonderful crea-tures? You have accelerated the pulsations of this world — my creatures — mechanical sons of my genius. You obey me — but you don't understand me. Why don't you understand me? (*pause*) Futility? Each one of my discoveries increases the bulk of this mountain of futility — now this mountain threat-ens to crush me. (*pause*) And I should keep working? No. I can't go on. And yet this minute machine is supposed to give back to the exhausted convicts the energy of youth. It is the machine that prolongs agony. (*does some tests*) I can't go on. To dehumanize mankind — to turn men into machines — machines that are not machines — men that are no longer men. (*pause*) And yet . . . maybe the last convulsion of this aberrant world will be born here? (*with savage joy*) After us — nobody else! The last inhabitants — the last tyrants! Then — why not speed it up? More fierceness — mutilations! This is Bacal's revenge! (*with rage*) Without remorse! New source of tortures — machine, you must work at once! Now! I want it! All the convicts must receive into their dry veins the electrons of my will — of my exasperation — of my in-junction to live! All of them — all will be thrown again into the furnace! (*he shivers; stares at the machines, toward which he stretches his fists*) Are you ready? Work — damn you — pulsate — live — kill . . . (*he writhes his mouth, totters, holds on to the machine, falls down dragging it with him, rolls under the bench*)

LIPA: (*appears on the threshold*) What a mysterious room! (*looks around*) A torture chamber! (*advances*) What are our enemies preparing? (*observes the machines on the bench with curiosity*) Machines — little machines — it's you that now love these insane men. And yet you're beautiful — I like you so much . . . (*ecstatic, she falls on the sofa*) Goddess Ma-chine — religion of the strong! Divine Machine — embrace me! Embrace me with your rounded, shining rods — your arms bend me in a red spasm. Let me kiss your belly, polished as that of an adolescent! My hair is caressed by your convulsed flywheels! Divine one — hug me — hold me tightly — take me entirely! I'll give you a mighty son — the winged son — the son-god! My beloved — for you only I feel myself woman! I feel you — I am entirely fused with you — burn me — ac-celerate the rhythm of your blazing pistons — submerge me with your life-giving oil . . . (*she writhes on the sofa*)

TONCHIR: (*getting up, stunned, and looking at the shattered machine*) On the floor . . . how? The machine — the avenging machine? (*totters*) I don't feel well — my head is about to explode . . .

LIPA: (*frightened*) Who's there?

TONCHIR: Who's speaking?

LIPA: A man?

TONCHIR: Who are you? My abode is forbidden to everybody. How dare you?

LIPA: I didn't mean to . . . Forgive me!

TONCHIR: Forgive? What does it mean? (*going up to her*) But you don't belong to our kingdom. Where do you come from? Were you trying to steal my secrets? Did you break my machine?

LIPA: I did you no harm. I am Lipa — your guest — your prisoner.

TONCHIR: Why did you come in here?

LIPA: I don't know. Certainly somebody directed me. Maybe you . . .

TONCHIR: Me?

LIPA: If not you . . . something that comes from you . . . somebody called me . . . drew me . . .

TONCHIR: I lost my strength. Fell down. Perhaps it is my perturbed soul that called you . . . But I cannot stand . . . (*he drags himself to the sofa*) Sit down. Thank you.

LIPA: Do you really want to spend your whole life in this diabolic laboratory — with all these mechanical pleasures?

TONCHIR: Pleasure no longer exists. There are stairs — one must go up — each new step gives new impetus — new courage. And the stairs are endless — one no longer feels the ascent — or the dizziness — one goes up — goes up like this — like this — without feeling anything.

LIPA: I don't understand your giving up life. You have changed the face of the world and of things. But you are not living. You are building ghosts . . .

TONCHIR: You are still young — still beautiful. But I am no longer a man . . .

LIPA: (*hugs him*) Don't say it . . . don't say it! (*kisses him*)

TONCHIR: I'm betraying my comrades.

LIPA: Your lips are cold. You gave me a light, metallic shiver. Your breath is a silver blade — which turns my passionate love to ice.

TONCHIR: My breath is a silver blade. It is the emanation of my metallized senses — which don't respond to the electricity of your radiant flesh. This is what we have become! Now I know what I still have to do . . .

LIPA: (*with joy*) Leave . . . abandon everybody . . .

TONCHIR: No! I stay. I'll punish myself for my mistake.

LIPA: Your mistake?

TONCHIR: . . . I wanted to be god!

LIPA: There's nothing to be punished for. Repudiate your mistake. Come back to life. (*pause*) The women are already sacrificed. Bacal spoke the last word. Save me! Save us! Only you can.

TONCHIR: Do you think so?

LIPA: You can do anything. Come back to the true life — stop — stop the machines — everything will collapse!

TONCHIR: No! Everything must stay as it is — I want to be punished for my mistake.

LIPA: Impossible! You cannot renounce your human nature . . .

TONCHIR: (*gets up*) You're wrong. I have given it up. Now I belong to this world. I am your enemy. I must be your enemy. Go. Farewell!

(*Lipa goes out slowly.*)

(*goes back to the bench and looks with distress at the shattered machine on the floor, then turns to the door*) You are gone — Lipa. You left in this cold place of computations — the glittering trace of your femininity . . . And here I am — old — useless . . .

BACAL: (*entering*) I need men — men — what about your machine?

TONCHIR: There — in pieces.

BACAL: Tonchir — what have you done?

TONCHIR: Me? . . . Nothing!

BACAL: Have you lost faith?

TONCHIR: Yes.

BACAL: Does the past oppress you?

TONCHIR: It's the present that paralyzes me.

BACAL: Come on! Let the sirens of your genius howl — make the earth tremble with the sinister violence of your machines. We must keep living — we're all inebriated with this life! Now everything is ours!

TONCHIR: I am a loser.

BACAL: What spiritual crisis is annihilating this ferocious creator?

TONCHIR: There is something stronger that we — you too know that you cannot defeat it.

BACAL: These are no longer Tonchir's words . . .

TONCHIR: I need to be left alone . . .

BACAL: (*going out*) I'll be watching your fighting back from a distance! I'll hear the howl of this vile soul — expelled by you!

(*The room gradually grows dark. From the background the Three Shadows emerge.*)

TONCHIR: My souls are calling me. These are the true souls — that I have eradicated from myself — in order to mold myself another one which is not mine. Now you appear before me — to remind me of what I used to be. But you can only fight over my corpse.

THE WHITE SHADOW: (*lit by an intense light, slowly comes forward*) I have never forsaken you — I still want you — take me with you — I beg you. Don't you see how pale I am — how wan. My breathing is agitated — it is fading. Why did you drive me away? I've been crying constantly — without tears my eyes — wilted — my eyes.

TONCHIR: I hardly understand you — hardly recognize you — and yet you have been mine — I dragged you with me — for so long — so long — so long . . .

THE WHITE SHADOW: Take me — I cannot die.

TONCHIR: Why do you want to upset me? I have no room for you any longer.

THE WHITE SHADOW: I am bound to you . . .

TONCHIR: You cradled my first youth. I enjoyed feeling your every pulsation — which instilled in me the supreme joy of life. But you made me a weak man. I found the strength to forget you.

THE WHITE SHADOW: A purple flower has bloomed on my mouth. This flower is reaching out with its tentacles — and my lips are completely white.

TONCHIR: (*stares at the Shadow; he is horrified*) How ugly — how unrecognizable my soul is! It's being bitten by the flower. The flower has disfigured it! (*shouting*) Get out! Out! Contaminated soul!

(*The White Shadow disappears.*)

My earthly souls — you have called me in order to increase my torture — to exasperate my last instants. No! I'm facing you. I have repudiated you. I repudiate you. I am no longer yours. I am Tonchir — the creator of all these machines which challenge the sky — the driving force of this kingdom hungry for planets. I am Tonchir! I am God! Hateful souls — do not persecute me any longer — I reject you — I curse you. (*he wanders about the room as a madman*)

THE RED SHADOW: (*comes forward*) Stop! Do not waver! Only I will give you oblivion.

TONCHIR: (*stops and looks at it in terror*) My red soul — voluptuous soul — my strong soul — do not take revenge, you too!

THE RED SHADOW: You are still in my suffocating coils.

TONCHIR: Don't be evil — I beg you.

THE RED SHADOW: I am the flame which burns and joins to the whole. Once I allowed you the most frenzied embraces — later I joined you to the machines. You created — dominated matter. I cannot be a stranger in you . . .

TONCHIR: The memory of you carries me away in a whirl into the flaming space — it consumes me like an asteroid. My body is saturated with all your joys. You are always in my blood. I don't want to shun you. I like the harm you do to me.

THE RED SHADOW: And now, would you let me consume myself in my devouring fire?

TONCHIR: Only you have given me the whole — and have annihilated me in the whole. I want to be rejoined with you — don't leave me — never again. (*he kneels at its feet and hugs its legs, laying them bare*) Don't leave me! (*he kisses its knees*)

THE RED SHADOW: Take me back with you — take me back . . .

TONCHIR: (*voluptuously, climbs up its body*) Your words do me so much good — they pick me up — they inebriate me — they blank me in ecstasy. I am no longer Tonchir, the violator of cosmic mysteries — I am the strong lover I used to be. And you must be still more beautiful — my violent soul — you are — more beautiful. Show me your face — your eyes — inextinguishable metallic flickering — your body — your entire body — irritating velvet dune — your pale ocher skin — Let me see your face — your face . . .

(*The Shadow unveils its face and shows its toothless skull.*

Page from The Broom, *October 1922. (New York Public Library, Picture Collection)*

and develops parallelly with the spirit of the times which seeks to contemplate, live and identify itself with reality itself.

THE AESTHETIC OF THE MACHINE AND MECHANICAL INTROSPECTION

We today, after having sung and exalted the suggestive inspirational force of *The Machine* - after having by means of the first plastic works of the new school fixed our plastic sensations and emotions, see now *the outlines of the new aesthetic of The Machine* appearing on the horizon like a fly wheel all fiery from Eternal Motion.

WE THEREFORE PROCLAIM

1. The Machine to be the tutelary symbol of the universal dynamism, potentially embodying in itself the essential elements of human creation : the discoverer of fresh developments in modern aesthetics.

2. The aesthetic virtues of the machine and the metaphysical meaning of its motions and movements constitute the new fount of inspiration for the evolution and development of contemporaneous plastic arts.

3. The plastic exaltation of *The Machine* and the mechanical elements must not be conceived in their exterior reality, that is in formal representations of the elements which make up *The Machine* itself, but rather in the plastic-mechanical analogy that *The Machine* suggests to us in connection with various spiritual realities.

4. The stylistic modifications of *Mechanical Art* arise from *The Machine-as-interferential-element.*

5. The machine marks the rhythm of human psychology and beats the time for our spiritual exaltations. Therefore it is inevitable and consequent to the evolution of the plastic arts of our day.

ENRICO PRAMPOLINI
(Translated by F. S.)

237

Wild cry from Tonchir. He falls on the sofa. The Red Shadow disappears.)

Who has broken my machine? The machine . . . Who's trying to take me . . . I'm suffocating . . . Bacal . . . men . . . all the men . . . the agony . . . the machine . . . do live! . . . pulsate . . . kill . . . !

(Music: repeat No. 2 from point C. From the windows bright red flashes: the intense life of the kingdom of the machines. Overwhelming music of the multiplied machines. Roars explosions whistles puffs grumbling of the engines. Pulsation of valves. Up and down of pistons. Shrieking of mechanical saws. Suddenly every noise stops. Pitch darkness. Then a beam of light cuts through the darkness aimed at the Black Shadow, which advances toward Tonchir, sits down next to him, puts its arms around him, and draws him near.)

THE BLACK SHADOW: I've been close to you — the only one close to you — the others wished to have you — but they were no longer the same — why didn't you call me?

TONCHIR: Fear — terror of you. Why did you come? Is it my hour?

THE BLACK SHADOW: I've been watching you all the time — you've always been my favorite — my son . . .

TONCHIR: I don't want you — I don't want to see you . . .

THE BLACK SHADOW: *(unveils its face)* I want to show myself to you only — I wander — I walk — wrapped in a black shroud — my wandering is restless — I go without peace. Everybody meets me — smiles at me — curses me. I bring down all the stairs that man has erected — I detour all the roads. To go where? Nobody knows. Truth is in the darkness. Walking in the darkness in order to find a glimmer of light — and the glimmer leads into another darkness. To wander — to wander — and never to find. And those who think they have found it — cannot discern it — because truth is God.

TONCHIR: Now — I am — in — your — power . . .

THE BLACK SHADOW: Get up! Come! (*disappears*)

(*Music: No. 4. Motif II. The stage lights up. Tonchir gets up, fearfully looks around. Then, as if attracted by a mysterious force, runs out.*)

III

The stage is dominated by the Machine-Brain, synthesis of the thought of the three despots: Bacal, Singar, and Tonchir. Men, convicts, and machines work according to the orders of this machine, which intercepts the leader's will. When the curtain rises, a group of Convicts Condemned to the Machines are rambling about the stage. Each wears a helmet, on top of which a little antenna glitters. They are moving in all directions; alternatively they run, stop, speed up again. Every movement is absolutely automatic. They look like inanimate beings, suspended in space and obeying invisible strings. Toward the end of the pantomime, Bacal and Singar enter. The Convicts gradually retreat. (Music: No. 5–I. Pantomime and Pandemonium.)

BACAL: Something is bending me — I try to stand up — it hurts . . .

SINGAR: Certainly it's not your age.

BACAL: My age is molded into platinum. Time cannot graze it — space cannot corrode it.

SINGAR: The burden of your thoughts?

BACAL: No. Somebody is clinging to my nerves. An implacable force . . .

SINGAR: Is it human?

BACAL: You said it.

SINGAR: Tonchir?

BACAL: You said it.

SINGAR: Tonchir is the executor.

BACAL: Tonchir is the driving force.

SINGAR: Tonchir is strong.

BACAL: I feel his strength. I have lost control. The screams of an intoxicated crowd are still carrying me away . . .

SINGAR: Lipa!

BACAL: Lipa is vengeance.

SINGAR: I'll kill vengeance.

BACAL: Too late.

SINGAR: You can do anything.

70

BACAL: Nobody will obey me.

SINGAR: Yours are the machines.

BACAL: The machines are shrieking.

SINGAR: Yours are the men.

BACAL: The men are laughing scornfully.

SINGAR: To give in and . . . destroy.

BACAL: To give in?

SINGAR: . . . and destroy!

BACAL: You too are failing me!

SINGAR: The first spark springs forth in the night — which muffles the rage of our pulsating machinery. They shake, exasperated — their oil bubbles. Thousands and thousands of sinister eyes anticipate the pleasure of burning lust . . . A thousand — one hundred thousand arms reach out — retreat — entwine convulsively. Everywhere mouths — mouths that seek each other — that roar kisses. The heart of the Machine-Brain accelerates its beats — still faster — at any moment it could break. Red clouds . . . Here the simun is coming — penetrates into our bones — melts us. Down there, our men are tearing each other to pieces — are killing each other — the corpses are piled upon each other in ever-rising heaps. Why? Why?

BACAL: No. It'll never happen! I still dare!

SINGAR: You no longer hold the command.

BACAL: I want Tonchir killed!

SINGAR: Your hand trembles! Tonchir's watching you!

BACAL: I will not kill — I don't want to kill Tonchir. He is too much a part of myself — of ourselves. I'll kill his thought — the sharp scythe that reaps everything.

SINGAR: His thought will live — even without his body.

BACAL: But we cannot go back. Tonchir is an obstacle. It's necessary to remove him.

SINGAR: How?

BACAL: I cannot take up a weapon. I can't . . .

SINGAR: Who can?

BACAL: The woman!

SINGAR: Do you think so? Would Lipa destroy her last hope? She claims her right to life! She deceives herself thinking that Tonchir . . .

BACAL: Lipa will share our life — she'll kill in herself that which still binds her to the Terrestrians.

SINGAR: You're insane! You don't know women. Lipa will not forget whom she's fighting for. Because what she's asking for others — she's asking primarily for herself.

BACAL: That's true! But Lipa is a foreigner. You'll force her . . .

SINGAR: (*goes up to the Machine-Brain*) This brain once so firm — so precise . . .

BACAL: Lipa!

SINGAR: This machine is not working anymore — it's sick. This is the desintegration of all your men — it's the end of your reign.

BACAL: Lipa! Tonchir!

SINGAR: Tonchir has elevated you!

BACAL: Now he longs for my end.

SINGAR: Can you kill a part of your Self?

BACAL: Since I have destroyed everything that was binding me to other men — in order to wander free and alone about my domain — since I have destroyed all feelings — reason for life — since I have destroyed love — inexorable violence of nature — I will cut off — even my arm — when this arm wants to rise up against me.

SINGAR: Intrepid raider of the soul's deserts — white lord of all the distances — more and more the fever of the unattainable is penetrating into your nerves. Now you're lying on the sand sparkling with crystal stars — your throat is dry — you have fallen — striken by the sun-vampire.

BACAL: Not I alone.

SINGAR: You'll be left alone. Your men have given you everything. You've taken everything. Now they are leering ghosts.

BACAL: My men will keep giving to me — in order to nourish the rhythm of life — which will never stop.

SINGAR: The machines are panting — they are exhausted. Their master is exhausted.

BACAL: Are you too conspiring?

SINGAR: Everything is bound to collapse when one raises oneself to excessive heights. You went beyond all barriers. You won. But in order to become God there is one last barrier. It's close to you. If you want to pass beyond it — you'll fall. I'm warning you. Stop! Not one more step — despot!

BACAL: Is it really you, the one in command of all my men? Is it you, my glorious comrade? There are no barriers. I dare everything. I want to be everything. Is it possible that you — Singar — want to leave me? No. Embrace me! (*they embrace each other*) Join your never-conquered soul with my own. Our kingdom is small. Let's join it to the cosmos! We'll win. We'll win again!

SINGAR: I've always followed you — I will follow you . . .

BACAL: To win is necessary! (*goes out*)

BOGO: (*entering*) The woman — the prisoner.

SINGAR: What about the condemned?

BOGO: They're still condemned!

SINGAR: At their places?

BOGO: Still at their places.

SINGAR: Tired?

BOGO: Entwined with the machines.

SINGAR: Are they happy with their fate?

BOGO: They couldn't be happier. They are machines!

SINGAR: Show her in!

(*Bogo goes out. Lipa enters.*)
You?

LIPA: At last! Here I am — my only wish was to be admitted to your presence.

SINGAR: I'm listening.

SINGAR: . . . but not as a woman.

SINGAR: How come?

LIPA: As a comrade.

SINGAR: You — a comrade? Then — you decided . . .

LIPA: Yes! But I demand . . .

SINGAR: I don't understand this word.

LIPA: I can demand. I do not belong to your kingdom.

SINGAR: You are our prisoner.

LIPA: You own my body . . . nothing else!

SINGAR: That — I think — is enough.

LIPA: I have decided to share your life. I want to give new birth to a man that you are lacking.

SINGAR: The man that we are lacking — already knows his way . . .

LIPA: If this man is no longer, you'll find yourselves defenseless against these new man-made giants.

SINGAR: Tonchir is no longer with us — he betrayed us.

LIPA: He has rebelled. The rebel is always great!

SINGAR: Do you love him?

LIPA: No! But I am all for him.

SINGAR: Does he want you?

LIPA: No.

SINGAR: Well then?

LIPA: I can no longer be alone.

SINGAR: There is no room here for you.

LIPA: I'll stay here — with you.

SINGAR: I'll cover you with chains.

LIPA: I'll stay with you!

SINGAR: Maybe — if you'll pass an ordeal — but this ordeal is terrible.

LIPA: I accept everything.

SINGAR: Don't say that.

LIPA: Anything.

SINGAR: Anything but this!

LIPA: Do you think that women are merely females?

SINGAR: This is to be seen. Do you really think that I am your enemy?

LIPA: I don't know. You certainly abuse your power.

SINGAR: What if I told you that this is painful to me?

LIPA: I wouldn't believe it.

SINGAR: Do you want to give Tonchir a proof of your love?

LIPA: What love? He doesn't know — he rejected me.

SINGAR: But I am still capable of understanding that. After all — why not? For what reason shouldn't you love Tonchir?

LIPA: Can it be you who speaks like this?

SINGAR: I am made the same way you are. I succeded in suffocating every rebellious cry of my self. I don't know yet whether this is a real strength. I'll never know.

LIPA: The doubt: the tip of a sword that grazes you. How are you going to parry it?

SINGAR: I let it penetrate all the way. I clench my teeth. I ignore the wound.

LIPA: Do you give up?

SINGAR: Not at all!

LIPA: Give me Tonchir!

SINGAR: Tonchir has reached the end of his existence — he wavers between life and death — he is again plagued by the ailments of mankind. Here we are inhuman! He who is human will not win. We have to be inhuman!

LIPA: I know. Nevertheless, this will not keep you from the hands of the gravedigger.

SINGAR: Does it matter? Those who devoted themselves completely to the agony of creation cannot surrender. The creator — the builder of a new world — cannot be a madman, a weak man. He is a supernatural being — aware only of his higher nature. How can you love someone who repudiates this greatness?

LIPA: I am defending the man he used to be.

SINGAR: You acknowledge his defeat.

LIPA: I am defending the man who returns.

SINGAR: Who returns to life or in the dust?

LIPA: Returning to his self.

SINGAR: I do love — I do love Tonchir. It's precisely this great love of mine that wants to suppress him.

LIPA: Aren't you aware of this cruelty?

SINGAR: If a man had succeeded in possessing your body — your soul — if he became your absolute master — you — for love — would give him everything — you would be the slave of his will. And what if this man — having become insane — prodigal — gave to somebody else — your body — your soul — these things which are yours — exclusively yours — and which you in a fever had given to him?

LIPA: Freedom! Death!

SINGAR: Your death or his? (*in hot pursuit*) Tonchir holds this kingdom in his hands — this entire precise world is regulated by his will. Now he's giving it up — he's surrendering it to sentiments — to mankind — to the enemy.

LIPA: (*subjugated by Singar's phosphorescent eyes*) Freedom! Freedom!

SINGAR: None of us can do it! You — who have been initiated into the

74

kingdom of the machines — yours is this glory! Strike! (*goes out*)

LIPA: I . . . strike . . . kill Tonchir? Never! Never! I'd rather kill myself . . .

(*Tonchir enters, dragging himself along with effort. His face is ghostly. He moves toward the Machine-Brain, sees Lipa, and stops as if upset. Then he goes up to her and gently touches her shoulder.*)

TONCHIR: Lipa!

LIPA: (*opens her arms*) Master!

TONCHIR: Are you happy in our kingdom?

LIPA: Are you kidding? I was supposed to kill you! . . .

TONCHIR: My comrades ordered you to do so — I know full well that you could never have done that. I anticipated everything . . .

LIPA: Is the spirit finally going to win?

TONCHIR: I've been able to see into my life. To be the only one — unique? I did it — because I felt it. It was necessary. Now I have gone beyond all that . . .

LIPA: You have created a world — you'll live for it!

TONCHIR: I am sacrificing myself for it . . .

LIPA: Your sacrifice is purposeless — it is harmful to everybody.

TONCHIR: I want to say good-bye for the last time to my creatures — to my great creature. I gave everybody a single brain — everybody thought alike — I have controlled their lives. I have killed individuality. This is my destruction! (*pointing to the Machine-Brain*) Now I'd like to remain alone with her . . . then . . . darkness . . . my soul will be gone.

(*Lipa goes out.*)

(*kneels down devoutly and kisses the Machine; his voice is broken by emotion*) My creature . . . my love . . . I'm going away forever. You are the winner — I the loser. Your soul is of metal — mine — unfortunately — is human!

(*The Machine answers by shrieking.*)

Now that I'm leaving you — I can understand how my insane dream has protected you — I can understand how much I have loved you. I gave myself entirely to you — you too frenziedly have offered me everything — lovers — lovers — tender — violent — cruel lovers.

(*The Machine shrieks louder and produces some sparks.*)

Stay with your masters — help them always — love them — love them as you loved me. Raise these proud men — higher and higher. (*waving good-bye with his hand*) Farewell! (*gets up and slowly goes out*)

(*The machine starts showing signs of insanity. Multicolored flames. The wires become red-hot and melt. It gurgles. Shrieks. Creaks. Crackles. Rattles.*)

BACAL: (*running in*) My men — my machines! Against whom are they rebelling? Everything is collapsing! Singar! Bogo!

(*Singar and Bogo run in; they are very pale.*)

Who dares move against me? I see a red cloud over there — which shrouds my empire.

SINGAR: The men have stopped — their mouths are twisted — their eyes open wide — they mumble — they are all gray — paralyzed.

BACAL: The machines? The convicts?

BOGO: The machines move — out of control — their hearts' precision has been broken. The convicts — look at them stupefied — they scream and laugh — they are out of their minds.

BACAL: Who has stopped us — whose hand is pushing us into nothingness? (*shouting*) Tonchir! Tonchir! Your soul is delirious! Tonchir! Call Tonchir!

BOGO: Tonchir is no longer among the living — he lies face down — his mouth kisses the earth . . .

BACAL: Tonchir used to adore a very pale female — shrouded by a black cape — her mouth too red — her face too pale . . .

(*Music: No. 6. Pantomime. The Convicts come in and mime. The Machine-Brain shivers, shrieks. Flashes of light become more and more intense. The Convicts move frenziedly, gesticulate, whimper.*)

My brain grows dim — the red cloud oppresses me — takes away my breath. (*runs toward the Machine*) The Machine is winning! We the creators — now the slaves! No! No! Machiiiiiiine! Maaaaaachiiiiiine! Stop (*throws himself onto the Machine as if to stop it*) You . . . you . . . are killing . . . everybody . . . everyyyyybodyyyyyy . . . everyyyyy . . .

(*A single flash of lightning. Bacal falls striken. The Machine stops. Immediately Singar, Bogo and the Convicts are petrified. The light fades gradually. Music: No. 7. Finale. The machines' sirens begin a painful, gloomy, agonizing song.*)

THE END

(Capri–Paris 1923)

MACHINE ART

Man and machine — which is Jekyll and which Hyde?
In the following pages the pens and brushes of many artists
provide an iconography for this ever-deepening
symbiotic relationship.

Les graveurs à la grec.

LEFT: *Nameless cogs in the relentless wheels of the burgeoning Industrial Revolution. The two engravers depicted in this ink and color wash from* Mascarade à la Grecque *by French artist Ennemond Alexandre Petitot (1727–1801) recall Shelley's lines from "Queen Mab":*

> *". . . obedience,*
> *Bane of all genius, virtue, freedom, truth,*
> *Makes slaves of men, and, of the human frame,*
> *A mechanized automaton."*

(The Metropolitan Museum of Art, The Elisha Whittelsey Fund, 1960)

BELOW: Locomotion *(n.d.), a hand-colored etching by Robert Seymour ("Shortshanks"), British, 1798–1836. Seymour's fantasy of future travel compares strikingly with the steam-age prophecy shown on page 145. (The Metropolitan Museum of Art, Gift of Paul Bird, Jr., 1962)*

OPPOSITE, ABOVE LEFT: Eléménts mécaniques *(oil, 1918–1923) by* Fernand Léger. *Perhaps more than any other twentieth-century artist, Léger — through his paintings and the memorable film* Ballet Mécanique *(1924) — explored the aesthetic potentialities of machinery. (Kunstmuseum, Basel)*

WALKING BY STEAM   RIDING BY STEAM   FLYING BY STEAM

Note ... In the Ladies Vehicle the Steam is made with a strong infusion of Gunpowder Tea   (*LOCOMOTION*.)   For an explanation of the Machinery see the next Number of the Edinburg Review.

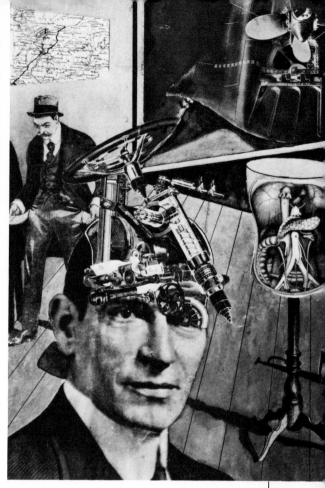

RIGHT: Tatlin at Home (1920) *pasted photo-engravings, gouache, and pen and ink by Raoul Haussman (b. Austria, 1886). (Moderna Museet, Stockholm. Copyright National-museum, Stockholm)*

BELOW: The Preparation of Glue from Bones (1920), *a collage by Max Ernst. The picture suggests a future development beyond Ray Bradbury's* Illustrated Man. *The* Illustrated Bionic Man, *perhaps? (Private Collection. Photograph courtesy Galerie André-François Petit, Paris)*

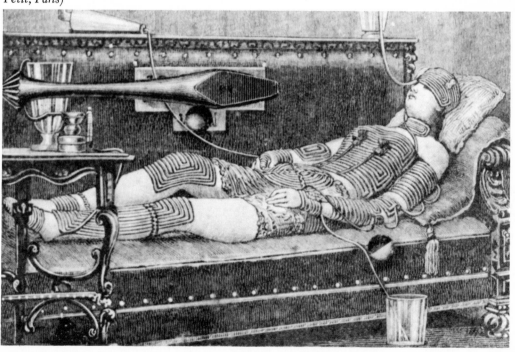

OPPOSITE ABOVE: *F. W. Seiwert's* Working Men *(oil, 1925) calls to mind this statement by Oscar Wilde: "On mechanical slavery, on the slavery of the machine, the future of the world depends."* — The Soul of a Man under Socialism *(Kunstmuseum, Düsseldorf. Photograph courtesy Landesbildstelle Rheinland)*

OPPOSITE BELOW: *Reductio ad automatum.* Man + Machine *(n.d.) by Bauhaus artist Kurt Schmidt. In this striking theater set design, stylized man and stylized robot are reduced to mere decorative elements in an environment that has, in turn, been reduced to the stark severity of a machine blueprint. (Reprinted from* The Bauhaus *by Hans M. Wingler, by permission of the M.I.T. Press, Cambridge, Massachusetts. English adaptation copyright © 1969 by The Massachusetts Institute of Technology. Courtesy Bauhaus-Archivs Berlin.)*

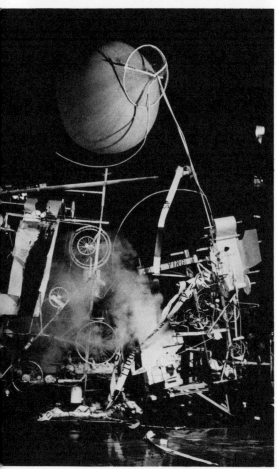

ABOVE: The Friendly Grey Computer — Star Gauge Model 54 *(1965) by Edward Kienholz. The artist's motorized construction consists of a rocking chair, doll's legs, instrument boxes, lights, switches, a metal case, a panel with numbers, a stack of index cards, a telephone receiver — and an instruction sheet. This last offers the following advice for operating Kienholz's wry response to our technology:*

Flashing yellow bulb indicates positive answer. Flashing blue bulb indicates negative answer. Green jewel button doesn't light so it will not indicate anything. Computers sometimes get fatigued and have nervous breakdowns, hence the chair for it to rest in. If you know your computer well, you can tell when it's tired and sort of blue and in a funky mood. If such a condition seems imminent, turn rocker switch on for ten or twenty minutes. Your computer will love it and work all the harder for you. Remember that if you treat your computer well it will treat you well.

(Collection, The Museum of Modern Art, New York. Gift of Jean and Howard Lipman)

BELOW: Homage to New York *(1960) by Jean Tinguely. This Dadaist work of machine-art — constructed from such "artifacts" as a typewriter, a Coke machine, several drums and fifteen motors — was designed to self-destruct — which it did, according to plan, on March 17, 1960, in the garden of the Museum of Modern Art. From the standpoint of 1977, Tinguely's "homage" to New York City seems to have been more like an ominous prophecy. (Copyright David Gahr, 1962)*

81

RIGHT: *An art deco technological robot benevolently embracing Man at the entrance of the Hall of Science at the Century of Progress 1933 World's Fair in Chicago. (Courtesy The Library, University of Illinois at Chicago Circle)*

BELOW: *In a manner reminiscent of the paint-by-numbers fad, a computer has turned a photo of a young woman into a series of numbers and back again. By measuring the levels of dark and light and then almost eliminating grays, the computer has turned the woman's face into what looks like a contour map. (Courtesy Bell Laboratories)*

OPPOSITE ABOVE: *Untitled photograph by Burk Uzzle. "The real cause for dread is not a machine turned human, but a human turned machine."*
— *Franz Winkler*
"The danger of the past was that men became slaves. The danger of the future is that men may become robots."
— *Erich Fromm*
(Copyright © Magnum Photos, Inc.)

OPPOSITE BELOW: *FM/Study 128 (1969) by Ernest Trova. Trova's silicon bronze gunman epitomizes the most destructive of all symbiotic relationships. (Courtesy The Pace Gallery, New York City)*

AUTOMATA:

PAST, PRESENT,

AND FUTURE

Maelzel's Chess-Player

by Edgar Allan Poe

Johann Nepomuk Maelzel (1772–1838), an Austrian acquaintance of Beethoven's, was credited with the invention of the metronome and of an automatic chess-player known as the Turk. In fact, the metronome was invented by Stöckel and the chess-player — which was not automatic at all — was actually the creation of Baron Wolfgang von Kempelen. The Turk's chess mastery was due to a concealed chess champion named William Schlumberger. When von Kempelen died in 1805, Maelzel acquired the Turk from the baron's son. Thereafter he toured Europe and the United States, staging tournaments between the chess-player and any and every challenger. The marvelous "machine" and its almost unbroken string of victories thrilled audiences on both continents, despite challenges such as the one reprinted in part on page 104. In the 1830s, while observing the Turk's performances in Boston, Edgar Allan Poe came to the conclusion that the so-called automatic chess-player could not possibly be a machine. His 1836 essay on the Turk displays the same logical incisiveness that characterized his pioneering flights into detective fiction. Maelzel's exhibit did not survive the 1830s. Poe's essay dealt a severe blow, but the *coup de grâce* was delivered in 1837 by a French journal, *Pittoresque*, which revealed that the whirring mechanism inside the Turk was mere fakery and that the so-called automatic chess-player concealed a human being. The moral, perhaps, was that in fact — if not in science fiction — it is better if what seems to be a robot does not turn out to be a man.

Perhaps no exhibition of the kind has ever elicited so general attention as the Chess-Player of Maelzel. Wherever seen it has been an object of intense curiosity to all persons who think. Yet the question of its *modus operandi* is still undetermined. Nothing has been written on this topic which can be considered as decisive, and accordingly we find everywhere men of mechanical genius, of great general acuteness, and discriminative understanding, who make no scruple in pronouncing the Automaton a *pure machine*, unconnected with human agency in its movements, and consequently, beyond all comparison, the most astonishing of the inventions of mankind. And such it would undoubtedly be, were they right in their supposition. Assuming this hypothesis, it would be grossly absurd to compare with the Chess-Player any simi-

Originally published in the *Southern Literary Messenger*, April 1836. The factual material in Poe's essay was based on information in Sir David Brewster's *Letters on Natural Magic Addressed to Sir Walter Scott* (1832).

lar thing of either modern or ancient days. Yet there have been many and wonderful automata. In Brewster's "Letters on Natural Magic," we have an account of the most remarkable. Among these may be mentioned, as having beyond doubt existed, firstly the coach invented by M. Camus for the amusement of Louis XIV. when a child. A table, about four feet square, was introduced into the room appropriated for the exhibition. Upon this table was placed a carriage six inches in length, made of wood, and drawn by two horses of the same material. One window being down, a lady was seen on the back seat. A coachman held the reins on the box, and a footman and page were in their places behind. M. Camus now touched a spring; whereupon the coachman smacked his whip, and the horses proceeded in a natural manner along the edge of the table, drawing after them the carriage. Having gone as far as possible in this direction, a sudden turn was made to the left, and the vehicle was driven at right angles to its former course, and still closely along the edge of the table. In this way the coach proceeded until it arrived opposite the chair of the young prince. It then stopped, the page descended and opened the door, the lady alighted, and presented a petition to her sovereign. She then re-entered. The page put up the steps, closed the door, and resumed his station. The coachman whipped his horses, and the carriage was driven back to its original position.

The magician of M. Maillardet is also worthy of notice. We copy the following account of it from the "Letters" before mentioned of Dr. Brewster, who derived his information principally from the "Edinburgh Encyclopædia," —

"One of the most popular pieces of mechanism, which we have seen, is the Magician constructed by M. Maillardet, for the purpose of answering certain given questions. A figure, dressed like a magician, appears seated at the bottom of a wall, holding a wand in one hand, and a book in the other. A number of questions, ready prepared, are inscribed on oval medallions, and the spectator takes any of these he chooses, and to which he wishes an answer, and, having placed it in a drawer ready to receive it, the drawer shuts with a spring till the answer is returned. The magician then arises from his seat, bows his head, describes circles with his wand, and, consulting the book as if in deep thought, he lifts it towards his face. Having thus appeared to ponder over the proposed question, he raises his wand, and striking with it the wall above his head, two folding-doors fly open, and display an appropriate answer to the question. The doors again close, the magician resumes his original position, and the drawer opens to return the medallion. There are twenty of these medallions, all containing different questions, to which the magician returns the most suitable and striking answers. The medallions are thin plates of brass, of an elliptical form, exactly resembling each other. Some of the medallions have a question inscribed on each side, both of which the magician answers in succession. If the drawer is shut without a medallion being put into it, the magician rises, consults his book, shakes his head, and resumes his seat. The folding-doors remain shut, and the drawer is returned empty. If two medallions are put into the drawer together, an answer is returned only to the lower one. When the machinery is wound up, the movements continue about an hour, during which time about fifty questions may be answered. The inventor stated that the means by which the different medallions acted upon the machinery, so as to produce the proper answers to the questions which they contained, were extremely simple."

The duck of Vaucanson was still more remarkable. It was of the size of life, and so perfect an imitation of the living animal that all the spectators were deceived. It executed, says Brewster, all the natural movements and gestures, it ate and drank with avidity, performed all the quick motions of the head and throat which are peculiar to the duck, and like it muddled the water which it drank with its bill. It produced also the sound of quacking in the most natural manner. In the anatomical structure the artist exhibited the highest skill. Every bone in the real duck had its representative in the automation, and its wings were anatomically exact. Every cavity, apophysis, and curvature was imitated, and each bone executed its proper movements. When corn was thrown down before it, the duck stretched out its neck to pick it up, swallowed, and digested it.[1]

But if these machines were ingenious, what shall we think of the calculating machine of Mr. Babbage? What shall we think of an engine of wood and metal which can not only compute astronomical and navigation tables to any given extent, but render the exactitude of its operations mathematically certain through its power of correcting its possible errors? What shall we think of a machine which can not only accomplish all this, but actually print off its elaborate results, when obtained, without the slightest intervention of the intellect of man? It will, perhaps, be said, in reply, that a machine such as we have described is altogether above comparison with the Chess-Player of Maelzel. By no means; it is altogether beneath it; that is to say, provided we assume (what should never for a moment be assumed) that the Chess-Player is a *pure machine*, and performs its operations without any immediate human agency. Arithmetical or algebraical calculations are, from their very nature, fixed and determinate. Certain data being given, certain results necessarily and inevitably follow. These results have dependence upon nothing, and are influenced by nothing but the data originally given. And the question to be solved proceeds, or should proceed, to its final determination, by a succesion of unerring steps liable to no change and subject to no modification. This being the case, we can without difficulty conceive the *possibility* of so arranging a piece of mechanism that, upon starting it in accordance with the data of the question to be solved, it should continue its movements regularly, progressively, and undeviatingly, towards the required solution, since these movements, however complex, are never imagined to be otherwise than finite and determinate. But the case is widely different with the Chess-Player. With him there is no determinate progression. No one move in chess necessarily follows upon any one other. From no particular disposition of the men at one period of a game can we predicate their disposition at a different period. Let us place the *first move* in a game of chess in juxta-position with the data of an algebraical question, and their great difference will be immediately perceived. From the latter — from the data — the second step of the question, dependent thereupon, inevitably follows. It is modelled by the data. It must be thus and not otherwise. But from the first move in the game of chess no especial second move follows of necessity. In the algebraical question, as it proceeds towards solution, the *certainty* of its operations remains altogether unimpaired. The second step having been a consequence of the data, the third step is equally a consequence of the second, the fourth of the third, the fifth of the fourth, and so on, *and not possibly otherwise*, to the end. But in proportion to the progress made in a game of chess, is the *uncertainty* of each ensuing move. A few moves having been made, no step is certain. Different spectators of the game would advise different moves. All is then dependent upon the variable judgment of the players. Now even granting (what should not be granted) that the movements of the Automaton Chess-Player were in themselves determinate, they would be necessarily interrupted and disarranged by the indeterminate will of his antagonist. There is then no analogy whatever between the operations of the Chess-Player and those of the calculating machine of Mr. Babbage, and if we choose to call the former a *pure machine* we must be prepared to admit that it is, beyond all comparison, the most wonderful of the inventions of mankind. Its original projector, however, Baron Kempelen, had no scruple in declaring it to be a "very ordinary piece of mechanism — a

---

[1] Under the head *Androides* in the "Edinburgh Encyclopædia" may be found a full account of the principal automata of ancient and modern times.

bagatelle whose effects appeared so marvellous only from the boldness of the conception, and the fortunate choice of the methods adopted for promoting the illusion." But it is needless to dwell upon this point. It is quite certain that the operations of the Automaton are regulated by *mind,* and by nothing else. Indeed, this matter is susceptible of a mathematical demonstration, *a priori.* The only question then is of the *manner* in which human agency is brought to bear. Before entering upon this subject it would be as well to give a brief history and description of the Chess-Player for the benefit of such of our readers as may never have had an opportunity of witnessing Mr. Maelzel's exhibition.

The Automaton Chess-Player was invented in 1769, by Baron Kempelen, a nobleman of Presburg, in Hungary, who afterwards disposed of it, together with the secret of its operations, to its present possessor. Soon after its completion it was exhibited in Presburg, Paris, Vienna, and other continental cities. In 1783 and 1784, it was taken to London by Mr. Maelzel. Of late years it has visited the principal towns in the United States. Wherever seen, the most intense curiosity was excited by its appearance, and numerous have been the attempts, by men of all classes, to fathom the mystery of its evolutions. The cut [p. 91] gives a tolerable representation of the figure as seen by the citizens of Richmond a few weeks ago. The right arm, however, should lie more at length upon the box, a chess-board should appear upon it, and the cushion should not be seen while the pipe is held. Some immaterial alterations have been made in the costume of the player since it came into the possession of Maelzel; the plume, for example, was not originally worn.

At the hour appointed for exhibition, a curtain is withdrawn, or folding-doors are thrown open, and the machine rolled to within about twelve feet of the nearest of the spectators, between whom and it (the machine) a rope is stretched. A figure is seen habited as a Turk, and seated, with its legs crossed, at a large box apparently of maple wood, which serves it as a table. The exhibiter will, if requested, roll the machine to any portion of the room, suffer it to remain altogether on any designated spot, or even shift its location repeatedly during the progress of a game. The bottom of the box is elevated considerably above the floor by means of the castors or brazen rollers on which it moves, a clear view of the surface immediately beneath the Automaton being thus afforded the spectators. The chair on which the figure sits is affixed permanently to the box. On the top of this latter is a chess-board, also permanently affixed. The right arm of the Chess-Player is extended at full length before him, at right angles with his body, and lying, in an apparently careless position, by the side of the board. The back of the hand is upwards. The board itself is eighteen inches square. The left arm of the figure is bent at the elbow, and in the left hand is a pipe. A green drapery conceals the back of the Turk, and falls partially over the front of both shoulders. To judge from the external appearance of the box, it is divided into five compartments — three cupboards of equal dimensions, and two drawers occupying that portion of the chest lying beneath the cupboards. The foregoing observations apply to the appearance of the Automaton upon its first introduction into the presence of the spectators.

Maelzel now informs the company that he will disclose to their view the mechanism of the machine. Taking from his pocket a bunch of keys he unlocks with one of them door marked 1 in the cut [p. 91], and throws the cupboard fully open to the inspection of all present. Its whole interior is apparently filled with wheels, pinions, levers, and other machinery, crowded very closely together, so that the eye can penetrate but a little distance into the mass. Leaving the door open to its full extent, he goes now round to the back of the box, and raising the drapery of the figure, opens another door

situated precisely in the rear of the one first opened. Holding a lighted candle at this door, and shifting the position of the whole machine repeatedly at the same time, a bright light is thrown entirely through the cupboard, which is now clearly seen to be full, completely full, of machinery. The spectators being satisfied of this fact, Maelzel closes the back door, locks it, takes the key from the lock, lets fall the drapery of the figure, and comes round to the front. The door marked 1, it will be remembered, is still open. The exhibiter now proceeds to open the drawer which lies beneath the cupboards at the bottom of the box — for although there are apparently two drawers, there is really only one — the two handles and two key-holes being intended merely for ornament. Having opened this drawer to its full extent, a small cushion, and a set of chess-men, fixed in a framework made to support them perpendicularly, are discovered. Leaving this drawer, as well as cupboard No. 1, open, Maelzel now unlocks door No. 2, and door No. 3, which are discovered to be folding-doors, opening into one and the same compartment. To the right of this compartment, however, (that is to say, the spectators' right), a small division, six inches wide, and filled with machinery, is partitioned off. The main compartment itself (in speaking of that portion of the box visible upon opening doors 2 and 3,

we shall always call it the main compartment) is lined with dark cloth, and contains no machinery whatever beyond two pieces of steel, quadrant-shaped, and situated one in each of the rear top corners of the compartment. A small protuberance about eight inches square, and also covered with dark cloth, lies on the floor of the compartment near the rear corner on the spectators' left hand. Leaving doors No. 2 and No. 3 open as well as the drawer, and door No. 1, the exhibitor now goes round to the back of the main compartment, and, unlocking another door there, displays clearly all the interior of the main compartment, by introducing a candle behind it and within it. The whole box being thus apparently disclosed to the scrutiny of the company, Maelzel, still leaving the doors and drawer open, rolls the Automaton entirely round, and exposes the back of the Turk by lifting up the drapery. A door about ten inches square is thrown open in the loins of the figure, and a smaller one also in the left thigh. The interior of the figure, as seen through these apertures, appears to be crowded with machinery. In general, every spectator is now thoroughly satisfied of having beheld and completely scrutinized, at one and the same time, every individual portion of the Automaton, and the idea of any person being concealed in the interior, during so complete an exhibition of that interior, if ever entertained, is immediately dismissed as preposterous in the extreme.

M. Maelzel, having rolled the machine back into its original position, now informs the company that the Automaton will play a game of chess with any one disposed to encounter him. This challenge being accepted, a small table is prepared for the antagonist, and placed close by the rope, but on the spectators' side of it, and so situated as not to prevent the company from obtaining a full view of the Automaton. From a drawer in this table is taken a set of chess-men, and Maelzel arranges them generally, but not always, with his own hands, on the chess-board, which consists merely of the usual number of squares painted upon the table. The antagonist having taken his seat, the exhibiter approaches the drawer of the box, and takes therefrom the cushion, which, after removing the pipe from the hand of the Automaton, he places under its left arm as a support. Then, taking also from the drawer the Automaton's set of chess-men, he arranges them upon the chess-board before the figure. He now proceeds to close the doors and to lock them — leaving the bunch of keys in door No. 1. He also closes the drawer, and, finally, winds up the machine, by applying a key to an aperture in the left end (the spectators' left) of the box. The game now commences — the Automaton taking the first move. The duration of the contest is usually limited to half an hour, but if it be not finished at the expiration of this period, and the antagonist still contend that he can beat the Automaton, M. Maelzel has seldom any objection to continue it. Not to weary the company is the ostensible, and no doubt the real, object of the limitation. It will of course be understood that when a move is made at his own table, by the antagonist, the corresponding move is made at the box of the Automaton, by Maelzel himself, who then acts as the representative of the antagonist. On the other hand, when the Turk moves, the corresponding move is made at the table of the antagonist, also by M. Maelzel, who then acts as the representative of the Automaton. In this manner it is necessary that the exhibiter should often pass from one table to the other. He also frequently goes in the rear of the figure to remove the chess-men which it has taken, and which it deposits, when taken, on the box to the left (to its own left) of the board. When the Automaton hesitates in relation to its move, the exhibiter is occasionally seen to place himself very near its right side, and to lay his hand now and then, in a careless manner, upon the box. He has also a peculiar shuffle with his feet, calculated to induce suspicion of collusion with

the machine in minds which are more cunning than sagacious. These peculiarities are, no doubt, mere mannerisms of M. Maelzel, or, if he is aware of them at all, he puts them in practice with a view of exciting in the spectators a false idea of the pure mechanism, in the Automaton.

The Turk plays with his left hand. All the movements of the arm are at right angles. In this manner, the hand (which is gloved and bent in a natural way), being brought directly above the piece to be moved, descends finally upon it, the fingers receiving it, in most cases, without difficulty. Occasionally, however, when the piece is not precisely in its proper situation, the Automaton fails in his attempts at seizing it. When this occurs, no second effort is made, but the arm continues its movement in the direction originally intended, precisely as if the piece were in the fingers. Having thus designated the spot whither the move should have been made, the arm returns to its cushion, and Maelzel performs the evolution which the Automaton pointed out. At every movement of the figure machinery is heard in motion. During the progress of the game, the figure now and then rolls its eyes, as if surveying the board, moves its head, and pronounces the word *"echec"* (check) when necessary.[2] If a false move be made by his antagonist, he raps briskly on the box with the fingers of his right hand, shakes his head roughly, and, replacing the piece falsely moved, in its former situation, assumes the next move himself. Upon beating and winning the game, he waves his head with an air of triumph, looks around complacently upon the spectators, and, drawing his left arm farther back than usual, suffers his fingers alone to rest upon the cushion. In general, the Turk is victorious; once or twice he has been beaten. The game being ended, Maelzel will again, if desired,

[2] The making the Turk pronounce the word *"echec"* is an improvement by M. Maelzel. When in possession of Baron Kempelen, the figure indicated a *check* by rapping on the box with his right hand.

exhibit the mechanism of the box, in the same manner as before. The machine is then rolled back, and a curtain hides it from the view of the company.

There have been many attempts at solving the mystery of the Automaton. The most general opinion in relation to it, an opinion, too, not unfrequently adopted by men who should have known better, was, as we have before said, that no immediate human agency was employed; in other words, that the machine was purely a machine and nothing else. Many, however, maintained that the exhibiter himself regulated the movements of the figure by mechanical means operating through the feet of the box. Others, again, spoke confidently of a magnet. Of the first of these opinions we shall say nothing at present more than we have already said. In relation to the second it is only necessary to repeat, what we have before stated, that the machine is rolled about on castors, and will, at the request of a spectator, be moved to and fro to any portion of the room, even during the progress of the game. The supposition of the magnet is also untenable; for, if a magnet were the agent, any other magnet in the pocket of a spectator would disarrange the entire mechanism. The exhibitor, however, will suffer the most powerful lodestone to remain even upon the box during the whole of the exhibition.

The first attempt at a written explanation of the secret, at least the first attempt of which we ourselves have any knowledge, was made in a large pamphlet printed at Paris in 1785. The author's hypothesis amounted to this — that a dwarf actuated the machine. This dwarf he supposed to conceal himself during the opening of the box by thrusting his legs into two hollow cylinders, which were represented to be (but which are not) among the machinery in the cupboard No. 1, while his body was out of the box entirely, and covered by the drapery of the Turk. When the doors were shut, the dwarf was enabled to bring his body within the box — the noise produced by some portion of the machinery allowing him to do so unheard, and also to close the door by which he entered. The interior of the Automaton being then exhibited, and no person discovered, the spectators, says the author of this pamphlet, are satisfied that no one is within any portion of the machine. This whole hypothesis was too obviously absurd to require comment or refutation, and accordingly we find that it attracted very little attention.

In 1789 a book was published at Dresden by M. I. F. Freyhere in which another endeavor was made to unravel the mystery. Mr. Freyhere's book was a pretty large one, and copiously illustrated by colored engravings. His supposition was that "a well-taught boy very thin and tall of his age (sufficiently so that he could be concealed in a drawer almost immediately under the chess-board)" played the game of chess and effected all the evolutions of the Automaton. This idea, although even more silly than that of the Parisian author, met with a better reception, and was in some measure believed to be the true solution of the wonder, until the inventor put an end to the discussion by suffering a close examination of the top of the box.

These bizarre attempts at explanation were followed by others equally bizarre. Of late years, however, an anonymous writer, by a course of reasoning exceedingly unphilosophical, has contrived to blunder upon a plausible solution — although we cannot consider it altogether the true one. His Essay was first published in a Baltimore weekly paper, was illustrated by cuts, and was entitled "An Attempt to analyze the Automaton Chess-Player of M. Maelzel." This Essay we suppose to have been the original of the *pamphlet* to which Sir David Brewster alludes in his "Letters on Natural Magic," and which he has no hesitation in declaring a thorough and satisfactory explanation. The *results* of the analysis are undoubtedly, in the main, just; but we can

Three views that expose the machinery — and fakery — of the Baron von Kempelen's automatic, chess-playing Turk. This "mechanical" prodigy checkmated Napoleon I but was no match for the deductive powers of Edgar Allan Poe, whose closely reasoned analyses of the Turk's chess-playing techniques proved conclusively that its intelligence could not possibly be that of a machine. (From J. F. Racknitz, Über den Schachspieler *[Leipzig, 1789], pls. 1–3. New York Public Library, Rare Book Division)*

only account for Brewster's pronouncing the Essay a thorough and satisfactory explanation, by supposing him to have bestowed upon it a very cursory and inattentive perusal. In the compendium of the Essay, made use of in the "Letters on Natural Magic," it is quite impossible to arrive at any distinct conclusion in regard to the adequacy or inadequacy of the analysis, on account of the gross misarrangement and deficiency of the letters of reference employed. The same fault is to be found in the "Attempt," etc., as we originally saw it. The solution consists in a series of minute explanations (accompanied by wood-cuts, the whole occupying many pages), in which the object is to show the *possibility* of *so shifting the partitions* of the box, as to allow a human being, concealed in the interior, to move portions of his body from one part of the box to another, during the exhibition of the mechanism — thus eluding the scrutiny of the spectators. There can be no doubt, as we have before observed, and as we will presently endeavor to show, that the principle, or rather the result of this solution, is the true one. Some person *is* concealed in the box during the whole time of exhibiting the interior. We object, however, to the whole verbose description of the *manner* in which the partitions are shifted, to accommodate

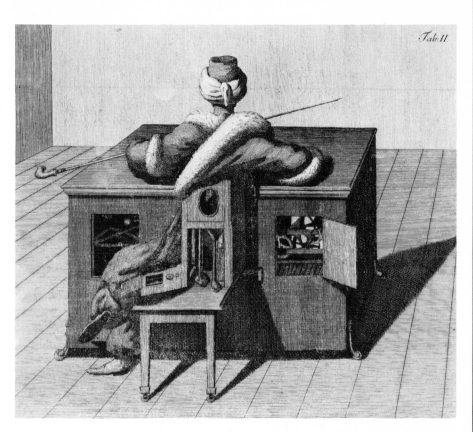

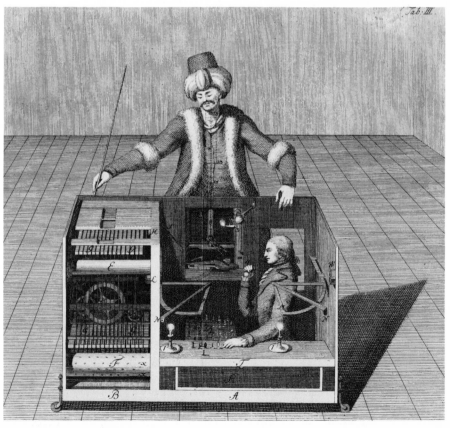

the movements of the person concealed. We object to it as a mere theory assumed in the first place, and to which circumstances are afterwards made to adapt themselves. It was not, and could not have been, arrived at by any inductive reasoning. In whatever way the shifting is managed, it is of course concealed at every step from observation. To show that certain movements might possibly be effected in a certain way, is very far from showing that they are actually so effected. There may be an infinity of other methods by which the same results may be obtained. The probability of the one assumed proving the correct one is then as unity to infinity. But, in reality, this particular point, the shifting of the partitions, is of no consequence whatever. It was altogether unnecessary to devote seven or eight pages for the purpose of proving what no one in his senses would deny — viz., that the wonderful mechanical genius of Baron Kempelen could invent the necessary means for shutting a door or slipping aside a panel, with a human agent too at his service in actual contact with the panel or the door, and the whole operation carried on, as the author of the Essay himself shows, and as we shall attempt to show more fully hereafter, entirely out of reach of the observation of the spectators.

In attempting ourselves an explanation of the Automaton, we will, in the first place, endeavor to show how its operations are effected, and afterwards describe, as briefly as possible, the nature of the *observations* from which we have deduced our result.

It will be necessary for a proper understanding of the subject, that we repeat here, in a few words, the routine adopted by the exhibiter in disclosing the interior of the box — a routine from which he *never* deviates in any material particular. In the first place he opens the door No. 1. Leaving this open, he goes round to the rear of the box, and opens a door precisely at the back of door No. 1. To this back door he holds a lighted candle. He then *closes the back door*, locks it, and, coming round to the front, opens the drawer to its full extent. This done, he opens the doors No. 2 and No. 3 (the folding-doors), and displays the interior of the main compartment. Leaving open the main compartment, the drawer, and the front door of cupboard No. 1, he now goes to the rear again, and throws open the back door of the main compartment. In shutting up the box no particular order is observed, except that the folding-doors are always closed before the drawer.

Now, let us suppose that, when the machine is first rolled into the presence of the spectators, a man is already within it. His body is situated behind the dense machinery in cupboard No. 1 (the rear portion of which machinery is so contrived as to slip *en masse* from the main compartment to the cupboard No. 1, as occasion may require), and his legs lie at full length in the main compartment. When Maelzel opens the door No. 1, the man within is not in any danger of discovery, for the keenest eye cannot penetrate more than about two inches into the darkness within. But the case is otherwise when the back door of the cupboard No. 1 is opened. A bright light then pervades the cupboard, and the body of the man would be discovered if it were there. But it is not. The putting the key in the lock of the back door was a signal on hearing which the person concealed brought his body forward to an angle as acute as possible — throwing it altogether, or nearly so, into the main compartment. This, however, is a painful position, and cannot be long maintained. Accordingly we find that Maelzel *closes the back door*. This being done, there is no reason why the body of the man may not resume its former situation — for the cupboard is again so dark as to defy scrutiny. The drawer is now opened, and the legs of the person within drop down behind it in the space it formerly occupied.[3] There is,

---

[3] Sir David Brewster supposes that there is always a large space behind this drawer even when shut; in

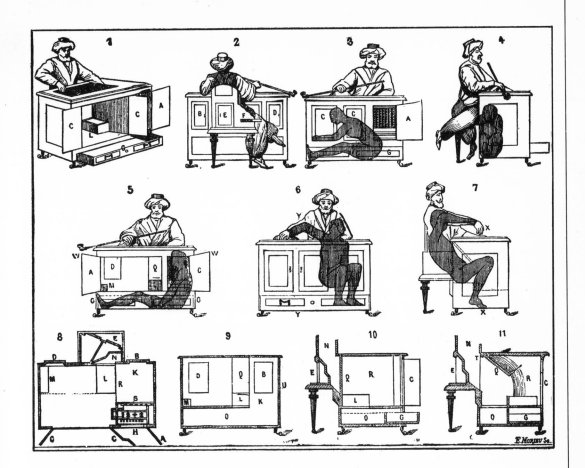

Another version of the workings of the automaton chess player, which is similar to Poe's description. (The Bettmann Archive, Inc.)

consequently, now no longer any part of the man in the main compartment — his body being behind the machinery in cupboard No. 1, and his legs in the space occupied by the drawer. The exhibiter, therefore, finds himself at liberty to display the main compartment. This he does — opening both its back and front doors — and no person is discovered. The spectators are now satisfied that the whole of the box is exposed to view — and exposed, too, all portions of it at one and the same time. But of course this is not the case. They neither see the space behind the drawer, nor the

interior of cupboard No. 1 — the front door of which latter the exhibiter virtually shuts, in shutting its back door. Maelzel, having now rolled the machine around, lifted up the drapery of the Turk, opened the doors in his back and thigh, and shown his trunk to be full of machinery, brings the whole back into its original position, and closes the doors. The man within is now at liberty to move about. He gets up into the body of the Turk just so high as to bring his eyes above the level of the chess-board. It is very probable that he seats himself upon the little square block or protuberance which is seen in a corner of the main compartment when the doors are open. In this position he sees the chess-board, through the bosom of the Turk, which is of gauze. Bringing his right arm across his breast, he actuates the little machinery necessary to guide the left arm and the fingers of the figure. This machinery is situated just beneath the left shoulder of the Turk, and is consequently

other words, that the drawer is a "false drawer," and does not extend to the back of the box. But the idea is altogether untenable. So common-place a trick would be immediately discovered — especially as the drawer is always opened to its full extent, and an opportunity thus offered of comparing its depth with that of the box.

97

easily reached by the right hand of the man concealed, if we suppose his right arm brought across the breast. The motions of the head and eyes, and of the right arm of the figure, as well as the sound "*echec*," are produced by other mechanism in the interior, and actuated at will by the man within. The whole of this mechanism — that is to say, all the mechanism essential to the machine — is most probably contained within the little cupboard (of about six inches in breadth) partitioned off at the right (the spectators' right) of the main compartment.

In this analysis of the operations of the Automaton, we have purposely avoided any allusion to the manner in which the partitions are shifted, and it will now be readily comprehended that this point is a matter of no importance, since, by mechanism within the ability of any common carpenter, it might be effected in an infinity of different ways, and since we have shown that, however performed, it is performed out of the view of the spectators. Our result is founded upon the following *observations* taken during frequent visits to the exhibition of Maelzel.[4]

1. The moves of the Turk are not made at regular intervals of time, but accommodate themselves to the moves of the antagonist — although this point (of regularity), so important in all kinds of mechanical contrivance, might have been readily brought about by limiting the time allowed for the moves of the antagonist. For example, if this limit were three minutes, the moves of the Automaton might be made at any given intervals longer than three minutes. The fact then of irregularity, when regularity might have been so easily attained, goes to prove that regularity is unimportant to the

[4] Some of these observations are intended merely to prove that the machine must be regulated *by mind*, and it may be thought a work of supererogation to advance farther arguments in support of what has been already fully decided. But our object is to convince, in especial, certain of our friends upon whom a train of suggestive reasoning will have more influence than the most positive *a priori* demonstration.

action of the Automaton; in other words, that the Automaton is not *a pure machine*.

2. When the Automaton is about to move a piece, a distinct motion is observable just beneath the left shoulder, and which motion agitates in a slight degree the drapery covering the front of the left shoulder. This motion invariably precedes, by about two seconds, the movement of the arm itself — and the arm never, in any instance, moves without this preparatory motion in the shoulder. Now let the antagonist move a piece, and let the corresponding move be made by Maelzel, as usual, upon the board of the Automaton. Then let the antagonist narrowly watch the Automaton, until he detect the preparatory motion in the shoulder. Immediately upon detecting this motion, and before the arm itself begins to move, let him withdraw his piece, as if perceiving an error in his manœuvre. It will then be seen that the movement of the arm, which, in all other cases, immediately succeeds the motion in the shoulder, is withheld — is not made — although Maelzel has not yet performed, on the board of the Automaton, any move corresponding to the withdrawal of the antagonist. In this case, that the Automaton was about to move is evident — and that he did not move was an effect plainly produced by the withdrawal of the antagonist, and without any intervention of Maelzel.

This fact fully proves: 1 — that the intervention of Maelzel, in performing the moves of the antagonist on the board of the Automaton, is not essential to the movements of the Automaton; 2 — that its movements are regulated by *mind* — by some person who sees the board of the antagonist; 3 — that its movements are not regulated by the mind of Maelzel, whose back was turned towards the antagonist at the withdrawal of his move.

3. The Automaton does not invariably win the game. Were the machine a pure machine this would not be the case; it would always win. The *principle* being dis-

covered by which a machine can be made to *play* a game of chess, an extension of the same principle would enable it to *win* a game; a farther extension would enable it to *win all* games, that is, to beat any possible game of an antagonist. A little consideration will convince any one that the difficulty of making a machine beat all games is not in the least degree greater, as regards the principle of the operations necessary, than that of making it beat a single game. If then we regard the Chess-Player as a machine, we must suppose (what is highly improbable) that its inventor preferred leaving it incomplete to perfecting it; a supposition rendered still more absurd, when we reflect that the leaving it incomplete would afford an argument against the possibility of its being a pure machine — the very argument we now adduce.

4. When the situation of the game is difficult or complex, we never perceive the Turk either shake his head or roll his eyes. It is only when his next move is obvious, or when the game is so circumstanced that to a man in the Automaton's place there would be no necessity for reflection. Now these peculiar movements of the head and eyes are movements customary with persons engaged in meditation, and the ingenious Baron Kempelen would have adapted these movements (were the machine a pure machine) to occasions proper for their display — that is, to occasions of complexity. But the reverse is seen to be the case, and this reverse applies precisely to our supposition of a man in the interior. When engaged in meditation about the game he has no time to think of setting in motion the mechanism of the Automaton by which are moved the head and the eyes. When the game, however, is obvious, he has time to look about him, and, accordingly, we see the head shake and the eyes roll.

5. When the machine is rolled round to allow the spectators an examination of the back of the Turk, and when his drapery is lifted up and the doors in the trunk and thigh thrown open, the interior of the trunk is seen to be crowded with machinery. In scrutinizing this machinery while the Automaton was in motion, that is to say, while the whole machine was moving on the castors, it appeared to us that certain portions of the mechanism changed their shape and position in a degree too great to be accounted for by the simple laws of perspective; and subsequent examinations convinced us that these undue alterations were attributable to mirrors in the interior of the trunk. The introduction of mirrors among the machinery could not have been intended to influence, in any degree, the machinery itself. Their operation, whatever that operation should prove to be, must necessarily have reference to the eye of the spectator. We at once concluded that these mirrors were so placed to multiply to the vision some few pieces of machinery within the trunk so as to give it the appearance of being crowded with mechanism. Now the direct inference from this is that the machine is not a pure machine. For if it were, the inventor, so far from wishing its mechanism to appear complex, and using deception for the purpose of giving it this appearance, would have been especially desirous of convincing those who witnessed his exhibition, of the *simplicity* of the means by which results so wonderful were brought about.

6. The external appearance, and, especially, the deportment of the Turk, are, when we consider them as imitations of *life*, but very indifferent imitations. The countenance evinces no ingenuity, and is surpassed, in its resemblance to the human face, by the very commonest of waxworks. The eyes roll unnaturally in the head, without any corresponding motions of the lids or brows. The arm, particularly, performs its operations in an exceedingly stiff, awkward, jerking, and rectangular manner. Now, all this is the result either of inability in Maelzel to do better, or of intentional neglect — accidental neglect being out of the question, when we consider that the whole time of the inge-

nious proprietor is occupied in the improvement of his machines. Most assuredly we must not refer the unlife-like appearances to inability, for all the rest of Maelzel's automata are evidence of his full ability to copy the motions and peculiarities of life with the most wonderful exactitude. The rope-dancers, for example, are inimitable. When the clown laughs, his lips, his eyes, his eyebrows, and eyelids — indeed, all the features of his countenance — are imbued with their appropriate expressions. In both him and his companion, every gesture is so entirely easy, and free from the semblance of artificiality, that, were it not for the diminutiveness of their size, and the fact of their being passed from one spectator to another previous to their exhibition on the rope, it would be difficult to convince any assemblage of persons that these wooden automata were not living creatures. We cannot, therefore, doubt M. Maelzel's ability, and we must necessarily suppose that he intentionally suffered his Chess-Player to remain the same artificial and unnatural figure which Baron Kempelen (no doubt also through design) originally made it. What this design was it is not difficult to conceive. Were the Automaton life-like in its motions, the spectator would be more apt to attribute its operations to their true cause (that is, to human agency within) than he is now, when the awkward and rectangular manœuvres convey the idea of pure and unaided mechanism.

7. When a short time previous to the commencement of the game, the Automaton is wound up by the exhibiter as usual, an ear in any degree accustomed to the sounds produced in winding up a system of machinery will not fail to discover, instantaneously, that the axis turned by the key in the box of the Chess-Player cannot possibly be connected with either a weight, a spring, or any system of machinery whatever. The inference here is the same as in our last observation. The winding up is inessential to the operations of the Automaton, and is performed with the design of excit-

ing in the spectators the false idea of mechanism.

8. When the question is demanded explicitly of Maelzel — "Is the Automaton a pure machine or not?" his reply is invariably the same — "I will say nothing about it." Now the notoriety of the Automaton, and the great curiosity it has everywhere excited, are owing more especially to the prevalent opinion that it *is* a pure machine, than to any other circumstance. Of course, then, it is the interest of the proprietor to represent it as a pure machine. And what more obvious and more effectual method could there be of impressing the spectators with this desired idea, than a positive and explicit declaration to that effect? On the other hand, what more obvious and effectual method could there be of exciting a disbelief in the Automaton's being a pure machine, than by withholding such explicit declaration? For, people will naturally reason thus:— It is Maelzel's interest to represent this thing a pure machine; he refuses to do so, directly, in words, although he does not scruple and is evidently anxious to do so indirectly by actions; were it actually what he wishes to represent it by actions, he would gladly avail himself of the more direct testimony of words; the inference is, that a consciousness of its *not* being a pure machine is the reason of his silence; his actions cannot implicate him in a falsehood — his words may.

9. When, in exhibiting the interior of the box, Maelzel has thrown open the door No. 1, and also the door immediately behind it, he holds a lighted candle at the back door (as mentioned above), and moves the entire machine to and fro with a view of convincing the company that the cupboard No. 1 is entirely filled with machinery. When the machine is thus moved about, it will be apparent, to any careful observer, that, whereas that portion of the machinery near the front door No. 1 is perfectly steady and unwavering, the portion farther within fluctuates, in a very slight degree, with the movements of the machine. This circum-

stance first aroused in us the suspicion that the more remote portion of the machinery was so arranged as to be easily slipped, *en masse*, from its position when occasion should require it. This occasion we have already stated to occur when the man concealed within brings his body into an erect position upon the closing of the back door.

10. Sir David Brewster states the figure of the Turk to be of the size of life; but in fact it is far above the ordinary size. Nothing is more easy than to err in our notions of magnitude. The body of the Automaton is generally insulated, and, having no means of immediately comparing it with any human form, we suffer ourselves to consider it as of ordinary dimensions. This mistake may, however, be corrected by observing the Chess-Player when, as is sometimes the case, the exhibiter approaches it. M. Maelzel, to be sure, is not very tall, but upon drawing near the machine, his head will be found at least eighteen inches below the head of the Turk, although the latter, it will be remembered, is in a sitting position.

11. The box behind which the Automaton is placed, is precisely three feet six inches long, two feet four inches deep, and two feet six inches high. These dimensions are fully sufficient for the accommodation of a man very much above the common size; and the main compartment alone is capable of holding any ordinary man in the position we have mentioned as assumed by the person concealed. As these are facts, which any one who doubts them may prove by actual calculation, we deem it unnecessary to dwell upon them. We will only suggest that, although the top of the box is apparently a board of about three inches in thickness, the spectator may satisfy himself, by stooping and looking up at it when the main compartment is open, that it is in reality very thin. The height of the drawer also will be misconceived by those who examine it in a cursory manner. There is a space of about three inches between the top of the drawer, as seen from the exterior, and the bottom of the cupboard — a space which must be included in the height of the drawer. These contrivances to make the room within the box appear less than it actually is, are referable to a design on the part of the inventor to impress the company again with a false idea, viz., that no human being can be accommodated within the box.

12. The interior of the main compartment is lined throughout with *cloth*. This cloth we suppose to have a twofold object. A portion of it may form, when tightly stretched, the only partitions which there is any necessity for removing during the changes of the man's position, viz., the partition between the rear of the main compartment and the rear of cupboard No. 1, and the partition between the main compartment and the space behind the drawer when open. If we imagine this to be the case, the difficulty of shifting the partitions vanishes at once, if indeed any such difficulty could be supposed under any circumstances to exist. The second object of the cloth is to deaden and render indistinct all sounds occasioned by the movements of the person within.

13. The antagonist (as we have before observed) is not suffered to play at the board of the Automaton, but is seated at some distance from the machine. The reason which, most probably, would be assigned for this circumstance, if the question were demanded, is that, were the antagonist otherwise situated, his person would intervene between the machine and the spectators and preclude the latter from a distinct view. But this difficulty might be easily obviated, either by elevating the seats of the company, or by turning the end of the box towards them during the game. The true cause of the restriction is, perhaps, very different. Were the antagonist seated in contact with the box, the secret would be liable to discovery, by his detecting, with the aid of a quick ear, the breathings of the man concealed.

14. Although M. Maelzel, in disclosing the interior of the machine, sometimes slightly deviates from the *routine* which we

e pointed out, yet *never* in any instance
es he *so* deviate from it as to interfere
ith our solution. For example, he has been
known to open first of all the drawer —
but he never opens the main compartment
without first closing the back door of cupboard No. 1; he never opens the main compartment without first pulling out the drawer; he never shuts the drawer without first
shutting the main compartment; he never
opens the back door of cupboard No. 1
while the main compartment is open; and
the game of chess is never commenced until
the whole machine is closed. Now, if it were
observed that *never, in any single instance*,
did M. Maelzel differ from the routine we
have pointed out as necessary to our solution, it would be one of the strongest possible arguments in corroboration of it; but
the argument becomes infinitely strength-
ened if we duly consider the circumstance
that he *does occasionally* deviate from the
routine, but never does *so* deviate as to
falsify the solution.

15. There are six candles on the board of
the Automaton during exhibition. The question naturally arises — "Why are so many
employed, when a single candle, or, at farthest, two, would have been amply sufficient
to afford the spectators a clear view of the
board, in a room otherwise so well lit up as
the exhibition room always is; when, moreover, if we suppose the machine a *pure
machine*, there can be no necessity for so
much light, or indeed any light at all, to
enable *it* to perform its operations; and
when, especially, only a single candle is
placed upon the table of the antagonist?"
The first and most obvious inference is, that
so strong a light is requisite to enable the
man within to see through the transparent
material (probably fine gauze) of which the
breast of the Turk is composed. But when
we consider the *arrangement* of the candles,
another reason immediately presents itself.
There are six lights (as we have said before)
in all. Three of these are on each side of
the figure. Those most remote from the
spectators are the longest, those in the mid-

dle are about two inches shorter, and those
nearest the company about two inches
shorter still; and the candles on one side
differ in height from the candles respectively
opposite on the other, by a ratio different
from two inches — that is to say, the longest
candle on one side is about three inches
shorter than the longest candle on the other,
and so on. Thus it will be seen that no
two of the candles are of the same height,
and thus also the difficulty of ascertaining
the *material* of the breast of the figure
(against which the light is especially directed)
is greatly augmented by the dazzling effect
of the complicated crossings of the rays —
crossings which are brought about by plac-
ing the centres of radiation all upon different
levels.

16. While the Chess-Player was in possession of Baron Kempelen, it was more than
once observed, first, that an Italian in the
suite of the Baron was never visible during
the playing of a game at chess by the Turk,
and, secondly, that the Italian being taken
seriously ill, the exhibition was suspended
until his recovery. This Italian professed a
total ignorance of the game of chess, although all others of the suite played well.
Similar observations have been made since
the Automaton has been purchased by Maelzel. There is a man, Schlumberger, who
attends him wherever he goes, but who has
no ostensible occupation other than that of
assisting in the packing and unpacking of
the Automaton. This man is about the
medium size, and has a remarkable stoop in
the shoulders. Whether he professes to play
chess, or not, we are not informed. It is quite
certain, however, that he is never to be seen
during the exhibition of the Chess-Player,
although frequently visible just before and
just after the exhibition. Moreover, some
years ago Maelzel visited Richmond with
his automata, and exhibited them, we believe, in the house now occupied by M.
Bossieux as a Dancing Academy. Schlumberger was suddenly taken ill, and during his
illness there was no exhibition of the Chess-
Player. These facts are well known to many

of our citizens. The reason assigned for the suspension of the Chess-Player's perform-ances was *not* the illness of Schlumberger. The inferences from all this we leave, with-out farther comment, to the reader.

17. The Turk plays with his *left* arm. A circumstance so remarkable cannot be acci-dental. Brewster takes no notice of it what-ever, beyond a mere statement, we believe, that such is the fact. The early writers of treatises on the Automaton seem not to have observed the matter at all, and have no refer-ence to it. The author of the pamphlet alluded to by Brewster mentions it, but acknowledges his inability to account for it. Yet it is obviously from such prominent dis-crepancies or incongruities as this that deductions are to be made (if made at all) which shall lead us to the truth.

The circumstance of the Automaton's playing with his left hand cannot have con-nection with the operations of the machine, considered merely as such. Any mechanical arrangement, which would cause the figure to move in any given manner the left arm, could, if reversed, cause it to move in the same manner the right. But these principles cannot be extended to the human organiza-tion, wherein there is a marked and radical difference in the construction, and, at all events, in the powers, of the right and left arms. Reflecting upon this latter fact, we naturally refer the incongruity noticeable in the Chess-Player to this peculiarity in the human organization. If so, we must imagine some *reversion,* for the Chess-Player plays precisely as a man *would not.* These ideas, once entertained, are sufficient of themselves to suggest the notion of a man in the inte-rior. A few more imperceptible steps lead us, finally, to the result. The Automaton plays with his left arm, because under no other circumstances could the man within play with his right — a desideratum, of course. Let us, for example, imagine the Automaton to play with his right arm. To reach the machinery which moves the arm, and which we have before explained to lie just beneath the shoulder, it would be necessary for the man within either to use his right arm in an exceedingly painful and awkward position (viz., brought up close to his body and tightly compressed between his body and the side of the Automaton), or else to use his left arm brought across his breast. In neither case could he act with the requisite ease or precision. On the contrary, the Automaton playing, as it actually does, with the left arm, all difficulties vanish. The right arm of the man within is brought across his breast, and his right fingers act, without any constraint, upon the machinery in the shoulder of the figure.

We do not believe that any reasonable objections can be urged against this solution of the Automaton Chess-Player.

The Automaton Chess-Player

Anonymous

. . . The machine, now exhibited by Mr. Maelzel, as an automaton, was made by M. de Kempelen, a Hungarian gentleman, and a member of the Aulic council of the German empire. . . .

The chess-player was made in the year 1769, and exhibited at Presburg, Vienna & Paris. In 1783, it was brought to London by the inventor, and exhibited during that, and the following year, to admiring multitudes.

The following account of it was published in Edinburgh, about fourteen years ago:

The chess-player of M. de Kempelen was a figure as large as life, dressed in a Turkish habit, and sitting behind a table with doors, three feet and a half in length, two in depth, and two and a half in height, and running on four wheels. The androides sits on a chair which is fixed to the table or commode: he leans his right arm on the table; in his left he holds a pipe, but with this arm he plays when the pipe is removed; a chess board of 18 inches being laid before him. The doors of the commode being thrown open, it is seen to contain wheels, levers, cylinders, and other pieces of mechanism; and in this state the machine is wheeled about the room. The vestments of the automaton are then lifted over its head, and the body is seen full of similar wheels and levers. A little door in its thigh is opened for a similar purpose; after which every thing being disposed in its place, the automaton is ready to play; and it always takes the first move.

From *The History and Analysis of the Supposed Automaton Chess Player of M. de Kempelen* (Boston: Hilliard, Gray, 1826).

At every motion the wheels are heard, the figure moves its head, and seems to look over every part of the chess board. When it checks the queen, it shakes its head twice, and thrice in giving check to the king. It likewise shakes its head when a false move is made, replaces the piece, and takes the move from the adversary. It generally, though not invariably, wins the game. M. de Kempelen or his substitute were always near the machine when it played, and wound it up like a watch, after it had made ten or twelve moves. A small square box was frequently consulted by the exhibitor during the game; and herein, he said, consisted the secret, which he could reveal in a moment. . . .

[A] curious little work, published at Paris, in the year 1785, . . . contains an explanation of the structure of the most celebrated automata of modern times. . . . According to this account, the machine was put in motion by a dwarf, a famous chess-player, who was concealed in the table, or commode. . . .

It remains to be explained, in what manner the dwarf hidden in the commode can know the game played by his adversary, and can direct the arm of the automaton at his pleasure. The most probable solution is, that the chessboard was made semi-transparent, so as effectually to conceal the person within, but to suffer the entrance of sufficient light for the dwarf to perceive whatever was done without. As for the means employed to give the necessary motions to the androides, little mechanical ingenuity is required. According to our authority, this was accomplished on the principles of the pantograph; an interior lever being governed by the dwarf, and made to describe a smaller circuit, while the arm of the machine described a similar circuit on an enlarged scale. . . .

Gonzalo Torres y Quevedo (left) matched with a chess-playing automaton designed by Norbert Wiener (right). Champion chess players can still hold their own against the skill of chess-playing computers, but the margin is narrowing. It is almost certain that some time within the next decade the world's chess champion will be a machine! If this development poses a threat to man's pride or prowess it does not seem to disturb the creators or enthusiasts of computerized games whose popularity is so rapidly proliferating. It may, in fact, be consoling for some individuals to envisage a period when man's only competitors may be machines rather than his fellow men. (M.I.T. Historical Collections)

The Strange World of M. Charliat

by John Kobler

In his tale "The Magic Shop" (*Twelve Stories and a Dream*, London, 1903), H. G. Wells mentions "a tiger in *papier-mâché* . . . that waggled his head in a methodical manner . . . a china hand holding magic cards, a stock of magic fishbowls in various sizes, and an immodest magic hat that shamelessly displayed its springs. On the floor were magic mirrors; one to draw you out long and thin, one to swell your head and vanish your legs, and one to make you short and fat like a draught. . . . A Magic Toy Sword. It neither bends, breaks, nor cuts the finger . . . magic trains that ran without steam and clockwork . . . and then some very, very valuable boxes of soldiers that all came alive directly you took the lid off. . . ." Such fantasy wonders are eclipsed by the real marvels in the Paris store of Georges Charliat who has devoted his life to collecting and restoring the complex automata of the eighteenth and nineteenth centuries. Through his skill and dedication, the "sublime toys" of the past are preserved for a present and future in which the functions of machines increasingly overshadow the original delight man discovered in creating or viewing automata that imitated life.

Among the antique dealers who thronged Koubbeh Palace in Alexandria last March to inspect the fabulous treasures of ex-King Farouk, before they were sold at auction by order of Egypt's new government, was a bouncy little Parisian of sixty namd Georges Charliat. Ignoring the solid gold and silver plate, the jewels as big as pheasant eggs, the centuries-old prayer rugs, Charliat marched straight up to a group of bizarre rarities known as automata, and for almost an hour stood rooted in rapt contemplation, his round, merry face aglow with wonder and delight.

Automata are essentially mechanical imitations of life. Wind-up walking bears and dolls that cry "mamma" are automata of sorts, though far too crude to command Charliat's attention for an instant. More typical of the wares which have earned him an international reputation among collectors of the rare and the curious is a troupe of five Arab circus performers, each about three inches high. Four of them squat cross-legged or stand around the trunk of a tree, banging and tootling on exotic music instruments, while the fifth, gripping a balancing pole, leaps up and down on a tightrope in a series of delicate, twinkling *entrechats*. This masterpiece was constructed more than 100 years ago by, or after the specifications of, Jean Eugène Robert Houdin, a celebrated French magician and one of the greatest designers of automata in the history of the craft.

Since the making of such wonder works

requires months or even years, their production is pretty much a thing of the past. The surviving specimens that still function perfectly fetch thousands of dollars in the antique market. For his Arab performers, Charliat recently turned down an offer of $5000.

Charliat is almost alone in the field. There are no more than three or four antiquaries in Europe dealing to any extensive degree in automata. In the United States the only ones are Hungarian-born Alexander Scheffer, of New York, whose Alavieille Russie, Inc., generally has some 100 assorted automata on hand, and the Berry-Hill brothers, Sidney and Henry, also New Yorkers, who confine themselves largely to pocket-size automata.

Of Farouk's automata, seventy-four in all, Charliat bid in seven, paying more than $10,000. The majority were in the form of pocket watches, but pocket watches the like of which few ordinary mortals have ever seen, let alone carried. "A very fine English eighteenth-century musical Automaton Repeater," runs a description of one in the catalogue prepared by the renowned London auction house of Christie, "the back brilliantly enameled with two women and a child in a market place, the back rises to disclose a harbour scene with a waterfall and moving boats in coloured gold." And all this, taking place within a frame less than two inches in diameter, did not begin to exhaust the effects of which the timepiece was capable. The Christie cataloguers, however, had been unable to describe the complete spectacle because the mechanism was partly damaged.

"Farouk's automata," Charliat reports, "were in almost as untidy a shape as his private life." Not until he got his prize watch back to his shop in Paris and tinkered with it for a month did he realize that one of the boats was a pirate craft which fired on the water front, setting a house afire, the smoke and flames being simulated by spinning bits of tinted glass. The watch in all its crazy glory is now available to anybody who wants to pay $3000 for it.

For the last seventeen years, ever since he began selling automata, Charliat has lived and worked at 46 Rue de Miromesnil, on the Right Bank, a short, narrow street of small specialty shops transecting one of the city's most distinctive shopping districts. The establishment on the right deals exclusively in handmade toy soldiers. The only other person in the place is his wife, a plump, brisk, strong-jawed woman, who keeps the books and runs up new clothes for time-worn male and female automata.

Of the wonderland within the shop the drab façade yields no hint. The dusty, dim-lit show window seldom contains anything more arresting than an antique jack-in-the-box, perhaps a pair of marionettes. The one room, narrow and low-ceilinged, can comfortably accommodate only two or three people at a time besides the Charliats. Even the automata themselves, which crowd the rest of the floor and wall space, do not normally add much excitement to the atmosphere, for Charliat keeps them inactive most of the time under glass bells or in cases, and refuses to put them through their paces merely to satisfy the curiosity of sight-seers, as he is constantly asked to do. On the rare occasions, however, when, to please a valued customer, he brings his favorite automata to life, the little shop becomes a Walt Disney-esque fantasy, a thousand children's dreams the night before Christmas.

On a velvet-draped bandstand, nine waxwork musicians, their eyes, heads and hands all in furious motion, launch into a concert of operatic medleys. In a bronze canary cage, a canary hops from one perch to another, chirping as he hops, turns around and hops back. A concealed rod moves him from perch to perch, but its action is so rapid, smooth and hair-fine precise that he seems to be flying unsupported through space. When Charliat acquired this amazing curio, it consisted of 150 separate parts packed in two shoe boxes, and he spent two months just figuring out how to reassemble them.

Music boxes tinkle, as Lilliputian pianists

run their fingers up and down keys like babies' teeth; clock dwellers emerge on the stroke of the hours to clash cymbals, ring bells, beat drums, march and caper; ballerinas pirouette atop powder boxes, their innards gently tickling and whirring.

Nobody is more enchanted by these sights and sounds than Charliat himself. "Georges," says Madame Charliat, "has never entirely left the world of childhood." The automata king himself says, "What I like is simple living and complicated mechanisms."

When asked to quote a price, Charliat seldom comes right out with it. This is not only guile but also a sincere reluctance to part with any automata. He loves them all. Few items in his shop ever go for less than $250, and the average figure runs between $1000 and $2000.

At those prices the clientele is of necessity limited. It has included the film star, Joseph Cotten, who collects circus clowns of every description and once bought a clown photographer from Charliat; Rossellini and Ingrid Bergman, who bought a singing bird in a cage; and an assortment of American and European millionaires who insist on anonymity.

Some pieces Charliat cannot bear to give up at any price. These, together with several he will sell, but which he considers too valuable to be stored at street level, he keeps in his five-room apartment above the shop, inaccessible to all but the choicest customers and fellow connoisseurs. The automaton which affords him the keenest delight, though it is by no means the most valuable, is an elephant made in 1761 by a Parisian craftsman named Pierre Gautier. All the creature can do is walk fifteen or twenty feet, but this it does with an incredible naturalness, its trunk swaying, its hide wrinkling and folding at the rear with the true elephantine falling-pants effect. Charliat never wearies of following it around the apartment.

Easily the most valuable piece Charliat owns is a nineteenth-century pistol with solid-gold chasings, a band of seed pearls around the muzzle and a watch in the butt. When the trigger is pressed, a singing bird pops up out of the barrel. It is one of four similar pistols made early in the nineteenth century by a pair of Geneva automatonists, the brothers Ami-Napoleon and Louis Rochat. Kaiser Wilhelm owned one, Sir David Salomons, a British collector, another, a Swiss dealer named Sandoz a third. The Frenchman who offered the fourth to Charliat, after inheriting it from his mother, could not get the bird to pop up, but he swore he had seen and heard it as a boy. Charliat decided to take a chance. Then for eight days he sweated over it. The bird would not emerge. Late at night on the ninth day he felt sure he had traced the trouble. He rushed into the bedroom to show his wife and pressed the trigger. The bird flew clear across the room. They spent the rest of the night on hands and knees, looking for it. Charliat keeps the pistol in a bank vault. It is insured for $10,000.

With almost every automaton Charliat buys he runs a risk. Few are in perfect working order, after passing from hand to hand for generations, and there is no certainty that they can ever be fixed. If not, Charliat's investment of time and money would be a dead loss. He banks confidently, however, on his long-practiced skill as a mechanic, even going so far as to issue a lifetime guarantee with every piece he sells. "If it worked once, it can be made to work again," he maintains. So far, his skill has never failed him, though some jobs have brought him to the verge of nervous collapse. Another automaton watch from Farouk's collection, which stood him $6000, occupied him day and night for six months.

The basic mechanism of most automata consists of sets of irregularly shaped cams strung on a camshaft. As the shaft revolves, each cam transmits its action through an endless chain to a cogwheel which transmits it to another cogwheel, and so on, until it reaches the part of the automaton to be animated. The more lifelike the total effect, the smaller and more numerous the cams and cogwheels. The simple lift of a dancer's leg may bring into play as many as thirty. The motive power is usually a watch spring.

Speech, music and other sounds involve additional complex devices. A combination of minuscule bellows and a whistle with a sliding piston, for example, produces the runs and trills of singing birds. The bellows, activated by a system of crank, connecting rod and lever, pumps air into the mouthpiece of the whistle. The piston forces it out through the aperture, producing the whistled notes.

When the entire apparatus is lodged in such objects as a lady's watch or the knob of a walking stick, as it often is, getting at a defective part is no work for butter fingers or restless temperaments. Charliat's powers of concentration are as intense as his fingers are deft. Once he has settled down to the workbench in the curtained-off alcove at the rear of his shop, a jewelers' eyepiece screwed into his eye, firecrackers might explode without distracting him. He has been known to bend over a handful of parts smaller than sequins for fifteen to twenty hours at a stretch without food, drink or rest, unconscious of the presence of friends or customers, deaf to the entreaties of his wife. "I am fascinated," he says, "by the prospect of resurrecting an automaton that hasn't functioned for a hundred years. It brings me an exhilarating sense of kinship with the artist who created it."

Charliat is continually devising special instruments for handling minute parts. He has screw drivers as thin as eyelashes, tweezers that can grip a speck of dust, needles fine enough to penetrate holes barely visible to the naked eye. When a part has to be replaced altogether, Charliat may fashion it himself on his own turret lathe or, if too elaborate for his equipment, turn the specifications over to a specialist. At the end of a long stint, Charliat usually has to lie down in a dark room for two days to rest his eyes.

Charliat literally fell into the automata business. While indulging in some horseplay twenty years ago he slipped while dancing, fracturing his leg in five places. The accident crippled him, interrupting a meteoric rise in the silent movies. Handsome hero of innumerable French cliff hangers, with a weekly income of around 8000 francs, in an era when the franc was twenty-five to the dollar, he had been hailed by one critic as "the bright hope of the French cinema." The hope remained unrealized. Though he recovered the normal use of his legs, he never went back to acting, devoting himself instead to automata. The choice was no second-best. In making it he fulfilled a boyhood dream.

At the age of six Charliat was already trying to animate various objects. He would partially stop up the mouth of the bathtub tap to increase the water pressure, insert an upward-curving length of pipe and on the crest of the resulting jet balance celluloid toys. It made a mess of the bathroom, but his parents, reluctant to discourage budding genius, did not interfere.

The parents, Georges Dmitri Charliat, of Russian origin, and his wife, Eugénie Delicie de Ligny, ran the swankiest flower shop in Paris, furnishing the floral decorations at all official receptions given by the French Government, the Russian and German embassies. They once journeyed to Russia to decorate Czar Nicholas II's yacht.

An important influence in Charliat's boyhood was his Uncle Alexandre, who headed an electrical trade school. From him he acquired a taste for mathematics and music, both of which are indispensable to a proper appreciation of automata. In fact, at the age of nineteen he decided to become a professional violinist.

He was studying the instrument in Vienna when World War I broke out and he landed in an internment camp. By the time he got home four years later, his father had died of influenza and his mother was having a hard struggle keeping up the florist shop by herself. He hastened to her aid, packaging potted plants by day, by night playing his fiddle in small concert halls.

One night, while he was entertaining at a private party, a woman guest, struck by his good looks, asked him how he would like to be in the movies. Upon his assurance that he would like it fine if it paid well, she told him she was a movie director [Alice Guy-Blaché?] and explained that what she had in mind for him was the lead in a six-installment serial titled *Gossette — Little Kid.*

With the money from this and many subsequent movie assignments, Charliat was able to indulge his undimmed passion for automata. His first important purchase was a grandfather clock that housed a dozen soldiers who marched out one door and back through another when the hour struck. He never remembers it without profound regret. It was the first piece he sold when he started out in the automata business.

When Charliat opened the shop on the Rue de Miromesnil in 1937, friends and family warned him that he would never find enough customers with both the taste and the pocketbook for such expensive oddments. They could not have been more mistaken. Within a year he had disposed of every automaton in his collection and was hunting for more. The only kind he cared to handle were about as common as red diamonds, so the hunt was arduous. It led him up and down Europe, ears to the ground for rumors of great odd estates in liquidation, of princely heirs willing to sell an heirloom or two, of wealthy homes with unexplored attics.

But sometimes the price would be too steep. For example, the owner wanted $500 for a Chinese clock enlivened by prancing figures, with which he fell in love in the south of France. Before Charliat could make up his mind, a private collector — a woman from Paris — grabbed it. Charliat never ceased to brood over his loss, cursing himself for an overcautious dolt. Immense was his joy when, five years later, the woman walked into his shop with the clock and

asked him to repair it. "Madame," said he, after ascertaining the extent of the damage, "the cost of repairs would be prohibitive. Why don't you sell it to me?" She did — for $1500.

As Charliat's reputation spread, he no longer had to bestir himself to replenish his stock. People with valuable specimens to sell would be likely to seek him out. Moreover, when word got around that he could repair the insanely intricate mechanisms, so many ailing automata were brought to him that in sheer self-defense he made it a rule, from which he has never departed, to work only on his own purchases.

Automata, as Charliat fondly points out, have a history much richer and older than most people realize. Like marionettes and puppets, they descend from the moving statues of ancient Egypt, and the earliest examples, millenniums old, had religious significance. These moving statues lacked internal mechanisms of any sort, but had articulated heads and limbs which could be manipulated by hand, like Charlie McCarthy. The leaders of religious cults used them for prophecy and in magic rites to awe their followers.

The first mechanized automata appeared during the third century B.C. in Alexandria, then the intellectual and commercial center of the world. An Alexandrian physicist named Hero constructed a series of them to dramatize various physical principles for his pupils. To show the motive force of steam, he mounted a group of little wooden figures on a turntable which he harnessed to the escape valve of a small water boiler. In doing so, he anticipated the steam engine by a couple of thousand years, though it never occurred to him or to any of his contemporaries to use it for industrial purposes.

With the development of clockmaking by the Arabs in the ninth century, the wheels-within-wheels, or Boob McNutt, era of automata was ushered in. Despite Mohammed's injunction against the representation of humans and animals in art, many of the early Arab clock automata were not only loaded with both but were displayed in public buildings. A celebrated one was the water-powered Elephant Clock at Fez, the work of the peerless clockmaker Al-Jazari, which had seven figures performing as many different actions, among them an eagle which dropped brass balls into the gaping mouth of a dragon which disgorged them into a vase, where they struck a gong, thus sounding the hours.

Europe's kings and nobles were quick to commission equally elaborate confections for their private amusement. Thomas Francine and his son, François, eminent hydraulic engineers from Florence, filled whole grottoes with mechanical grotesqueries on the grounds of Henri IV's castle in St.-Germain-en-Laye, near Paris. "The Persian grotto was very spacious," to cite a few details among hundreds from a contemporary description. "Perseus, bigger than life size, descended from aloft and with a sword struck an enormous dragon emerging from a fountain; the dragon instantly sank to the bottom. In the distance one perceived Andromeda, fastened by chains to a great rock, and the chains mysteriously unfastened themselves. Opposite, a figure representing Bacchus sat on a barrel atop a rock, holding a glass in his hand. From both glass and barrel issued streams of water which, as they reached the ground, set in motion little blacksmiths, weavers, carpenters, knife grinders and other artisans, each working at his trade." But the feature which Henri IV, an incurable practical joker, enjoyed most was a goddess who suddenly squirted water at people when they ventured too close.

The golden age of automata was the eighteenth century, with its worship of science and reason. The idea of reproducing natural phenomena by mechanical means exerted a peculiar fascination upon the minds of the day, and when exhibited publicly, the "sublime toys," as they were sometimes called, drew enormous crowds. Frequently the credulous would mistake appearance for reality and attribute to an automaton a mysterious life of its own.

In England, one Henry Bridges completed, after twenty-two years of nerve-racking effort, a multifarious contraption he called "the Microcosm or the Universe in Little," and toured Europe and America with it. *The New York Mercury* reported that it showed "(1) all the celestial phenomena, (2) the nine muses giving a concert, (3) Orpheus in the forest, (4) a woodworking studio, (5) a delicious grove, (6) a beautiful countryside with view of the sea." The mechanism included 1200 cogwheels.

In Neuchâtel, Switzerland, Pierre Jaquet-Droz and his son, Henri-Louis [*sic*], created a number of what they termed "androids" — from the Greek *andros*, man, and *eidos*, form. The Neuchâtel museum acquired three of their masterpieces in 1906 for 75,000 Swiss francs — then equivalent to $15,000. They are exhibited publicly nowadays only on the first Sunday of the month. The first is a moon-faced lad, almost three feet high, who draws the heads of Louis XVI and Marie Antoinette. The second, a girl, plays a spinet. The third another lad, sits behind a writing desk, pen in hand. In a recent typical demonstration, he dipped his pen in an inkwell, shook it twice and wrote, neatly forming each letter and with due attention to dotted *i*'s, crossed *t*'s and spacing: *"Soyez les bienvenus à Neuchâtel"* — Welcome to Neuchâtel. He is capable of writing any sentence, as the letter-forming apparatus can be reset as easily as a typesetting machine.

The Franklin Institute, in Philadelphia, has a girl android, made by an associate of Jaquet-Droz, Henri Maillardet, which draws as well as writes. Charliat once sold for $1500 a near-life-size automaton poet who fluttered his eyes and sighed as he penned a sonnet.

But the most prodigious automaton ever was a duck. In 1735 there arrived in Paris from his native Grenoble a twenty-six-year-old mechanical wizard named Jacques de Vaucanson, whose ambition was no less than to create life artificially. To raise money for experiments, he determined to construct some automata that people would pay well to see. Three years later he introduced a duck with a copper body which quacked, bathed itself, drank water, ate grain, digested it and voided.

Few live performers ever enjoyed greater success. But Vaucanson presently got bored with his brain child and sold it to two promoters who exhibited it all over Europe. From them it passed through numerous ownerships, winding up in the private collection of an eccentric German doctor, Godfrey Christopher Beireis, of Helmstadt. During a visit to Beireis' house in 1805, Goethe and his son saw the duck, or what was left of it. "Defeathered and reduced to a skeletal state," the German Shakespeare wrote in his journal, "the duck still ate his oats heartily, but no longer digested them."

For the next thirty-five years the whereabouts of Vaucanson's duck remained unknown, though every automata dealer in the world tried to trace it. The operator of a traveling museum of automata, a Swiss named Johann Bartholomew Reichsteiner, finally stumbled on the legendary creature in Berlin. A clockmaker with whom he was chatting chanced to remark that an extraordinary duck had been deposited with him as security for a loan by a Hungarian, a Professor Döbler, who then disappeared. Satisfying himself that the hocked duck was indeed Vaucanson's, the Swiss set out in search of the owner. Months later he found him in Prague and persuaded him to sell out.

But the duck proved to be in such bad shape through disuse that repairing it practically amounted to creating a new one. Involving more than 4000 parts, it kept Reichsteiner busy for three and a half years and cost him approximately $3350. The reborn animal made its debut on the stage of Milan's world-famed La Scala opera house in 1844 and scored as huge a success as the original.

"It is the most admirable thing imaginable," a member of the audience recalled later, "an almost inexplicable human achievement. Each feather of the wings is mobile. . . . The artist touches a feather on the upper portion of the body, and the bird lifts its head, glances about, shakes its tail feathers, stretches itself, unfolds its wings, ruffles them and lets out a cry, absolutely natural, as if it were about to take flight. The effect is still more astonishing when the bird, leaning over its dish, begins to swallow the grain with incredibly realistic movement. As for the method of digestion, nobody can explain it."

Except in the most general way, nobody ever managed to discover how any part of it worked. Robert Houdin advanced the theory that the stomach contained chemicals which reduced the grain to liquid. Encouraged by the triumph of the reconstructed duck, Reichsteiner spent three years creating a second one.

What happened to the two ducks is a mystery as tantalizing to automata lovers as is the disappearance of Judge Crater to criminologists. Reichsteiner, it is known, eventually went broke in litigation over an invention of his. According to a contemporary biography, he stored the first duck in a barn. The second he gave to a Leipzig café proprietor in 1865, probably in settlement of a debt, and the café proprietor sold it to somebody called Bernus. Neither duck has been seen since.

Not long ago, however, the curator of Paris' Museum of Arts and Crafts unearthed in a storeroom some photographs left behind by his predecessor, bearing the caption: "Views of Vaucanson's duck, received from Dresden." There are four of them. They all show a skeleton, stripped of all plumage, perched on a platform underneath which can be discerned a tangle of cams, wheels and motors. The photographs give no hint as to where and when they were taken, nor whether the duck is Vaucanson's or Reichsteiner's. That somewhere in the world one or both ducks may still exist, however, is a possibility to which Charliat fondly clings.

The thought of it sometimes comes to him in the dead of night, and when it does, he can sleep no more for wondering how they might be recovered and restored to their pristine glory.

Although the production of "sublime toys," like a Vaucanson duck or a Robert Houdin tightrope walker, is an art no longer practiced, automata of a different kind and purpose, the lineal descendants of Heron's practical demonstrations, flourish in the modern world. They are the electronically controlled robots and "thinking machines" which perform innumerable tasks for business and industry. The classic androids imitated man's physical appearance and actions; the twentieth-century automata, which look like nothing human, imitate his intellectual processes.

But for Georges Charliat, the happy tinkerer, the small boy at heart, who in spirit inhabits another time and place, these scientific miracles hold little interest. The most versatile robot the electronic laboratories ever created does not fascinate him nearly so much as Pierre Gautier's little elephant, born 1761, as it pads, trunk swinging and bottom baggy, across his living-room floor.

Silent Sam: a traffic control automaton. Since the 1960s in the U.S., live flagmen have been increasingly replaced by such robots as Silent Sam — a battery-operated, six-foot-high automaton who can tirelessly control traffic twenty-four hours a day in all weather conditions. Silent Sam is extensively used by metropolitan police departments, electric light and power utilities, turnpike and bridge authorities, telephone line crews, road construction contractors, airports, guard rail and striping contractors, and highway lighting firms. (Queens Devices, Inc., Long Island City, N.Y.)

ʃilent ʃam

by Susan Sheehan

In the right circumstances, no human being minds taking orders from an automaton, and the most diehard union members have no objection to robots being given the kind of job that can be boring as well as dangerous. So there seems to be a rosy future ahead for Silent Sam. . . .

The other day, as we were driving across the Triborough Bridge, we spotted a cheerful, energetic-looking figure waving a bright-red flag to warn us of some road-repair work up ahead. The figure, dressed in a bright-blue shirt and trousers, a red helmet and vest, and brown work shoes, was standing on a red-and-white striped pedestal and seemed altogether human until we got within twenty feet of him, when we realized that he was a robot. Back in our office, prompted by thoughts of the robot as the ideal construction-site employee (no wages, vacations, or fringe benefits; no sassy comments to pretty female motorists; no complaints about long hours or inclement weather), we phoned the Triborough Bridge and Tunnel Authority and asked what they could tell us about the robot. "Yeah. Silent Sam. We've got two of them, and they seem very effective," the man at the T.B. & T.A. said, and then he gave us the name and phone number of Daniel Berne, the general manager of Queens Devices, Inc., Silent Sam's manufacturer. We called Mr. Berne and accepted an invitation to come out to his office, in Long Island City, for a chat and an even closer look at the first robot on the T.B. & T.A.'s payroll. Mr. Berne turned out to be a Brooklyn-born, wavy-haired, soft-spoken man in his fifties. Silent Sam turned out to have pleasant features (honest brown eyes, strong nose, quiet grin), fashionably long sideburns, and an olive complexion. "That's the color the first model came out in, and we decided it was a good way to keep it," Mr. Berne said. "Not black, not white — racially acceptable. He could be Latin-American. With his weather-bronzed face, he represents the typical outdoor worker." Silent Sam appears to be in his late twenties or early thirties, and he looks healthy. He is six feet tall, but, thanks to his pedestal, which is eighteen inches high, and to larger-than-life proportions (broad shoulders, muscular arms), he seems taller than that. His price, nine hundred dollars, includes a D.C. motor but not a twelve-volt battery (thirty-five dollars), which is what keeps his right arm waving — at a choice of speeds — for between five and seven days at a clip. He is made of heavy-duty plastic, and his vest, flag, helmet, and base are painted with Day-Glo paint, which fluoresces when light hits it.

We asked Mr. Berne who had thought up Silent Sam, and when, and why, and he told us that the responsible party was Martin Kaltman, a friend of his and one of the owners of the Queens Lithographing Corporation. "Marty was driving to my house

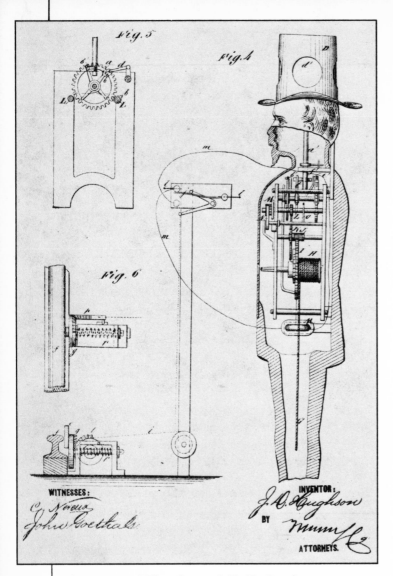

A precursor of Silent Sam — J. D. Hughson's electric railroad signal of 1877. (The Bettmann Archive, Inc.)

one night back in early '67," Mr. Berne said. "He was annoyed by the poor way traffic was being handled at a construction site between his house and mine. There were no flagmen on duty. Marty knew that several flagmen were killed by cars and trucks each year, and he thought that a three-dimensional mechanical flagman might be a good idea."

Mr. Kaltman turned his idea over to Mr. Berne, and Queens Devices, Inc., was launched. Two years, three artists' sketches, and several models and motors later, Silent Sam in his present incarnation (earlier versions had "sloppy clothes" and an impracti-

cal A.C. motor) was ready for the marketplace. "In early '69, we exhibited Sam in Chicago at the Construction Equipment Convention and received tremendous raves," Mr. Berne said. "But then we ran into the realities of life. Before anyone could use Silent Sam in a particular state, the commissioner of highways of that state had to give his approval. We started working on getting approvals from the commissioners of fifteen states, and you can imagine the red tape, the lethargy, and the time involved. Meanwhile, we realized that there was another market for Silent Sam — the independent authorities, like the Triborough Bridge and Tunnel Authority, that don't come under any state's supervision or control. The Port of New York Authority has a Silent Sam, and the Pennsylvania Turnpike Commission is using two, and we're staging demonstrations across the country. Silent Sam was tested in Indiana by radar engineers and was found to be more effective in slowing down traffic than flagmen, signs, or a combination of flagmen and signs. We've sold just a handful so far, but we expect to be in mass production very soon."

We asked Mr. Berne whether the unions had indicated any opposition to Silent Sam, and he said they hadn't because they were able to fill only twenty-five per cent of the available flagman jobs. "No one wants to stand in rain and snow with cars and trucks whizzing by him at sixty miles an hour," Mr. Berne told us. "It's a dull, dangerous job. Sam is popular with his co-workers. He makes them feel safer. Flagmen take coffee breaks; Sam doesn't. Even if his battery runs down and his arm stops moving, he will still give good protection just standing there. A few months ago, on a rainy night, there was a bad fire in Mamaroneck. The Fire Department knew a man who works in our company who lives there, and asked to borrow Silent Sam. Sam replaced three policemen who had been directing traffic. Our man ran into one of the cops in a coffee shop across the street. 'If that guy is stupid enough to stand there in the rain and wave his flag while I'm drinking hot coffee, let him do it,' the cop said."

Automaton, or the Future of the Mechanical Man

by H. Stafford Hatfield

The following pages are excerpted from a work that is of interest primarily for its summation of the problems and prospects of robotization on the eve of the development of automation and the computer. After surveying the range of automatic principles and devices in use during the twenties, the writer suggests a number of ways in which automata could be used to ameliorate the conditions of modern industry, transport, and domestic life. Noticing that "little or nothing" has been done to dispense with "the dreary, monotonous, and disgusting labor inseparable from industry as carried on today," he deplores the turning of modern workers into human automata. The essay concludes by urging a truly humanitarian application of the potentialities of the machine and looks forward to an age in which real automata will liberate human automata to live rich, creative, individual lives.

. . . From the dawn of . . . civilization . . . our poets and writers have embodied our aspirations in legends of man-made automata. . . . Frankenstein and his "monster" enjoyed a great vogue in the eighteenth century [sic]. Recent examples are the Robot play and the film "Metropolis". The Greek legend of Pygmalion, similar superficially, in reality illustrates how different was the Greek point of view. Pygmalion was an artist, not a scientist. He was not trying to produce an automaton to keep house for him, but to embody in a figure his ideal of beauty. It lived only when a soul was breathed into it in response to his prayers. In our legends, the demoniacal men of science who make the automata often receive back-handers from the artists who write the legends. The automata acquire souls and thus pass beyond the control and comprehension of the men of science. In our day,

however, this artist's way of ending the legend makes us smile. We feel that the soul would be a nuisance both to ourselves and to the unfortunate automaton, whereas we know that a practicable Robot would satisfy a long-felt need. . . .

The Future of the Automaton. — The purpose of this essay is to consider how far we may reasonably expect [the] development of an automatic link between instrument and machine to go in the future. We have . . . considered the matter from the point of view of the human worker, and we have concluded that he will not improve in dependability. On the other hand, it is certain that the automaton can be largely improved and cheapened. Also, that the processes of the future will be more exacting from the point of view of control than those of the present. Hence it is likely that the only limits which will be set to the development of automata will be those inherent in the limitations of our science. . . .

From H. Stafford Hatfield, *Automaton, or the Future of the Mechanical Man* (London: Routledge & Kegan Paul Ltd., 1928). Reprinted by permission.

The Mechanical Brain

Examples of Simple Automata. — We have . . . defined the automaton as consisting of three parts, corresponding to the senses, brain, and limbs of the human body. It is thus distinguished from an automatic tool, such as a lathe, which is devoid of senses, and is thus incapable of adapting itself to varying conditions of material or working. Perhaps the simplest everyday example of a true automaton is the gramophone motor. Its function is to turn the disc at a constant speed, in spite of varying friction, strength of spring, and other causes. In it we find a very simple instrument, consisting of a pair of spring governor balls, the divergence of which is a measure of the speed of the motor. This device "perceives" any change in speed, from whatever cause. It acts upon a brake applied to the motor, presses this on when the speed increases, and takes it off when it decreases. This is the brain of the arrangement, which controls the limbs, the motor. . . .

The Corrector Automaton. — [Such] devices are typical of one fundamental kind of automaton. It is a kind which may be said to wait for slight trouble, and then immediately take steps to correct it. . . .

[It] is possible to construct automata which do not wait for trouble to occur, but, instead, anticipate it. Such automata perceive a change in conditions, such as size or composition of material, and immediately adjust the tool so as to meet correctly the changed conditions. As an example we may take the case of softening water by the addition of chemicals which throw down the lime and magnesium salts to which the hardness is due. The dose of chemical added must be exactly proportioned to the hardness of the water. Now the process of softening and settling takes two hours to complete, so if we were to try and work on the "corrector" principle, and have our automaton readjust the dose of chemical when it found that the treated water was too hard or overdosed,

we should always be two hours late, and the device would work with very great inaccuracy. Instead, we use an instrument (invented by the writer) which analyses the raw water as it enters the plant, and at once adjusts the valves to give the correct dose of chemical. The automatic analyst exactly imitates the usual method by which liquids are analysed by hand in the laboratory, and can thus be used for many other purposes besides the analysis of water. Innumerable technical processes carried out on natural materials, such as ores, wood fibre, vegetables (sugar beet), and so forth, require at present continual control by analysis, since the raw material varies in a way which cannot be foreseen. These processes will no doubt be rendered largely automatic in the future.

The . . . diagram illustrates this "prescient" type of automaton. The electrical scheme is used merely because it is simple, to exhibit the principle. It is evident that this . . . type of automaton has inherent advantages which make it of wide application. Broadly speaking, whenever we can construct an instrument to measure the quality to which our tool or plant has to be adjusted we can construct an automaton of this type.

Sorting. — Perhaps one of the simplest cases of this kind is that of sorting. The scope for future development is here very great, and the problems to be solved are very various. It is of importance to consider it at this stage, because it illustrates very well a matter of fundamental importance to our subject. The problem is clearly one for the "prescient" type of automaton, which should perceive the size, weight, or other quality according to which sorting is to take place, and direct the objects to be sorted into their appropriate receptacles.

Now in the matter of size, the problem is solved in the most complete fashion by one of the oldest devices known to man, namely, the sieve. A series of closely graded sieves

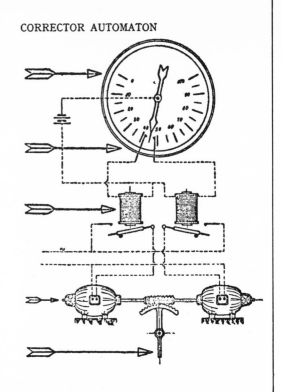

SENSE.　　　　　　　　　INSTRUMENT
(such as pyrometer, pressure gauge, speed indicator, compass) measuring whatever the automaton is to keep constant (temperature, pressure, speed, direction).

BRAIN.
　　　　　　　TWO FIXED CONTACTS
close together, one on either side of instrument pointer. They are set so that when pointer is touching neither, its reading is the correct constant value desired. The two contacts are connected to a battery and to

　　　　　　　TWO RELAY SWITCHES
one of these receives a weak current from the battery, whenever instrument pointer touches one of the contacts: that is, whenever that which should be constant varies in either direction. It then closes and holds on a switch which sets going one of

LIMBS.　　　　　　　　TWO MOTORS
As soon as error is | which then moves
rectified, instrument |
pointer ceases to |
touch contact, relay |
switch ceases to re- |
ceive current, and | **REGULATING LEVER**
lets motor switch go. | in correct direction to
Motor stops. | rectify error.

placed one above the other in order of fineness will, when set vibrating, separate a pulverized mass thrown upon the uppermost into as many classes of fineness as we may desire. The various products may be passed on to machines each adjusted to treat material of the particular size. The treatment of ores in mining makes use of this and numerous other classifying devices. The actual ore-dressing machines, which separate the valuable from the worthless material are also closely allied in function. These devices, classifiers, and dressing plant are automata of the most ideal prescient description. But they can hardly be said to be built upon the human tripartite model. They perform their tasks at a vastly greater speed and lower cost than could devices measuring each individual grain, and actuating mechanism in accordance with the result to send that grain to its appropriate destination. They warn us that there may be a way round the anthropomorphic solution of any automaton problem.

Hand-Picking. — Yet in this very field there is a problem which has hitherto resisted all attempts at solution. In treating minerals, it is often very desirable for many reasons to avoid having to crush up the rock containing the mineral to the point at which all the constituents are set free from one another. Often the valuable material occurs in large lumps sparsely distributed through barren rock, and to crush the whole rock would mean heavy expense and great waste. So instead, what is called "hand-picking" is resorted to. The rock is broken into large lumps, which pass on a travelling belt between rows of workers who pick out by eye the lumps containing valuable material in visibly large quantity. The same process is, of course, resorted to in numerous other cases where good material is to be sorted from bad.

Leaving aside for a moment the question of whether we shall ever succeed in making an instrumental eye to work actually in the same way as the human worker does, it is evident that solutions of the problem are possible in cases where we can satisfactorily perceive what is to be perceived by means other than eye. For instance it should not be difficult to make an automaton to sort lumps of rock or other objects in accordance with their magnetic qualities, weight, or electric conductivity.

It would be impossible to go into details here, but there can be no doubt that a very great development will take place in this field. An immense amount of sorting and counting is done by hand which could be done automatically. Each problem, however, is quite individual. We will therefore content ourselves with considering quite generally the possibilities of automatic sorting or recognition by means analogous to the human eye.

The Seeing Automaton. — Ever since the discovery of the sensitivity of the element selenium to light, which causes it to change its electrical resistance very greatly, inventors have promised themselves the most marvellous results by an application of its properties. It was early proposed to switch street lighting lamps on and off automatically as required by means of a selenium cell which would "perceive" the amount of natural light, and operate the switch accordingly. A recent invention is Mr. Fournier d'Albe's apparatus by which the blind are enabled to read ordinary print. . . . We may safely prophesy that the day is at last at hand when a perfectly satisfactory and highly sensitive "light-relay" will be available; an apparatus, that is to say, that will respond to light falling upon it by good strong variations in an electric current, which can then be set to work any kind of apparatus.

Possibilities of the Seeing Automaton. — The writer is, indeed, puzzled to know why such an apparatus is not already on the market. Its applications would be innumerable. The old scheme of turning artificial light on and off automatically according to need could at once be put into practice. Races could be automatically timed by sending a beam of light across the track into a light-sensitive relay, which would start or stop a chronograph when horse, man, car, greyhound, or other competitor interrupted the beam for a moment. Objects of any sort passing a given point could be counted by similar means. The advent of vehicles at cross-roads could be signalled in advance.

Many of these things could be done, by the way, by apparatus much less delicate than a light-relay, since the beam of light could be powerful enough to actuate a relay sensitive to its heating effect. . . .

Limitations to the Powers of the Electric Eye. — Let us now consider the limitations of our electric eye. We may assume that there is no limit to its sensitiveness, and that it could be made to react with certainty to the minutest variations in light of any selected colour. The limitations we are about to consider are of a different description, and raise one of the most fundamental questions of our subject. Very low forms of life possess a faculty of responding to visual impressions which depends, not upon an extreme sensitiveness of the eye, but upon the comparison of the impression received at the moment with past impressions stored in the memory, the latter word being used in the widest sense, of a "conditioned reflex". Once an animal has experienced what food, or an enemy, looks like, it afterwards responds to the visual impression in a reflex, that is, a purely automatic fashion. Now this faculty has a further peculiarity, namely, that it is not limited to recognition of an impression exactly like that which originated the reflex, but is equally well exercised by similar impressions. An animal responds by instant flight to an impression of human beings of the most varied sizes, shapes, and colours. It appears to have a general idea of what men are like, and is able to compare a visual impression with this general idea (not always correctly, of course) and judge whether it has been received from a human being. . . .

How far can we ever hope to imitate this in our automata? Could we ever, for instance, devise an apparatus which would signal "sail ho!" at sea?

Let us proceed step by step. Fournier d'Albe has given us, in his reading machine for the blind, one hint of a possible solution. A tiny camera serves as its eye, taking in one letter at a time, and projecting an image of that letter upon a screen. The screen, like our own retina, is a network of separate light-sensitive spots. In his machine, it is true, the impression on the artificial retina is converted into sound in a telephone applied to the ear of the blind reader. Each spot on the retina plays a certain note when it is excited by light, and the reader thus learns to distinguish the particular mixture of musical sound which corresponds to each letter. We could, however, equally well arrange that each sensitive spot, when excited, should transmit the effect to a set of indicators on a board corresponding to the retina. Say the image of a white cross fell upon the instrument. Then, let us say, studs lifted by little magnets would rise up on the board to form a sort of image of the cross. How is our automaton to be made to "know" that this is a cross, and to do whatever we want it to do when it receives the image of a cross? Obviously, if all we ask is that it should respond to a particular size of cross set the same way every time, the matter is quite simple. The same set of studs would rise every time, and we have only to arrange that that particular set shall complete an electric circuit. But if we demand a response to any cross, no matter what size, but to no other shape; or worse still, to a cross of any size in any position, the problem will need some hard thinking. We will return to it later, and see whether we can imagine some solution.

We may safely say, however, that we could devise a machine which would typewrite, or set type, from a typewritten or printed page presented to it. It would be highly complicated and expensive, but it could be done. It would be in the nature of an adaptation of the blind reader.

The Automaton which Reads Manuscript. — Suppose, however, the ingenious constructor were now asked to adapt his machine to read manuscript, or even type of different size and "face". The difficulty would be stupendous; I should say insuperable. Our automaton has to "recognize," not a given thing of fixed size and shape, but the class to which a vast variety of similar sizes and shapes belongs.

Automatic Traffic Control. — Another example is the problem of automatic traffic control. A good light-sensitive relay would, as I have already remarked, readily enable a signal to be sent that *something* was approaching a cross road. It would be next door to impossible, however, to signal "man," "car," "horse and cart," let alone "policeman," "Ford," "Rolls-Royce." It is not impossible, however, that we could imitate the sensitiveness of men and animals to movement in the field of vision. We might even be able to signal "object moving rapidly", and like matter, but it would be very difficult.

It appears to be worth considering whether this problem of the automatic signalling of traffic might not be approached from quite a different side. All the vehicles the advent of which requires to be signalled are made of magnetic material, and it would be a perfectly easy matter to arrange that the passage of such a vehicle at a point some distance from the cross roads should cause a signal to be exhibited there, which would be washed out again at the moment when the vehicle reached the cross road. The difficulty would be to avoid the complete disorganization of the system by vehicles not behaving in a regular fashion.

The Ear of the Automaton. — We pass naturally from sight to sound. A brilliant achievement in this field was the development of sound ranging during the war. As with the eye, so with the ear; our device will respond to a certain sound by sending or varying an electric current. Given such a response, and we can do almost anything

"We'll All Be Happy Then," from Life, *1911.*
(New York Public Library, Picture Collection)

with it by modern methods. Recent inventors have been busy with devices which respond selectively to certain notes of the musical scale. Others make use of sound waves in water and other media. As sound is propagated at quite a slow rate it is possible to measure distances automatically by sending out sound waves and noting the time they take to reach their destination, or to be reflected from it and return to the sender. Very successful devices of this kind enable ships to take soundings to any depth.

It is extremely likely that sound-perceiving instruments will, in future, play a part in the construction of automata. They may well serve as means for detecting the nature of media, whether solid, liquid, or gaseous. For instance, the pitch of the note of a whistle depends upon the nature of the gas used to blow it. This fact was used by Haber to construct an extremely simple means for indicating the presence of dangerous gases in the air of mines. . . .

The Automaton that Types to Dictation. — Every now and then we hear from the inventor who plans a typewriter that will write to dictation. This may appear to the non-technical no more fabulous than a forecast of the achievements of wireless would have appeared twenty years ago. But the cases are not on all fours. I venture to say that if any technical problem can be flatly termed insoluble, this is one. We are brought back at the start to the same type of difficulty as that discussed in connection with the electric eye, namely, of the "recognition" by our automaton of similarity, and not identity. The voice and pronunciation of every person differs very greatly. Were we to succeed, however, in making an electric ear which would respond in the same way every time to the same sound, it would still respond only to syllables, and not to letters. Its only conceivable response would be a sort of pattern corresponding to each syllable, which would then have to be sorted automatically amongst not 26, but thousands of possibilities. Finally, of course,

the matter of spelling would present entirely insuperable difficulties, since the human typist can only decide by the sense whether to write "w o o d" or "w o u l d," and often decides wrongly.

This matter, however, takes on a somewhat different aspect when considered in connection with the endeavours, so persistent and yet hitherto so unsuccessful, towards the formation of a universal language. The latest suggestion of the kind, however, appears to me to have far more prospect of success, and, curiously enough, to reopen the question of the typist automaton. I mean the plan for the simplification of English proposed by Mr. C. K. Ogden. He hopes that when his present researches are completed it will be possible, by a new and fundamental analysis of the mechanism of language, to produce a modified English with a basic vocabulary of about 500 words in which almost everything can be expressed. It is certainly within the bounds of possibility that an automaton might be devised to take dictation in this language from a practised speaker, provided only that no ambiguities of sound occur in the final vocabulary.

The recording of sound upon the photographic film, now in commercial use in the talking film, should have a future in connection with office work. It would enable a record, secret if desired, to be made of important conferences or interviews. It would be a much more capacious and handy apparatus than the dictaphone. It could, of course, be used to avoid typing altogether by sending records of spoken messages by post; an early dream of the future, which was common when the phonograph was first invented. It should also find a wide application in the construction of automatic announcing devices, such as would be very useful to railways.

The Recognition Problem Again.— It is not at all impossible that light might be thrown upon some of the fundamental problems in psychology by close study of recognition. We can only go so far here as to discuss visual recognition a little more in detail, and leave the reader to consider it further for himself. A lens throws an image of a shape upon a screen made up of light-sensitive spots. Can we arrange matters so that when and only when, a certain shape appears on the screen a certain response is given? The answer is certainly, "Yes," if the shape is the same size and occupies the same place every time. Can we further provide the apparatus with a repertoire of shapes, so that it gives a separate signal for each shape in its repertory? The answer is "Yes," if the same shape always occupies the same part of the screen. Suppose it does not? It is quite conceivable that the apparatus might then be made to centre itself automatically upon, say, the centre of figure of the shape, whatever it was. It would get a general disturbance, "figure in the field," and then proceed to move until it was "looking" at the centre of the figure. The image of the latter might then, however, be set at the wrong angle for recognition, so that the next step of the automaton would be to turn the screen right round the compass. If recognition still failed, it might be because the figure was too large or too small. It would then be necessary to arrange that the automaton would change the size of the image on the screen (by optical means) in minute gradations, and test out each gradation all round the compass. It would no doubt require a lifetime of work and a fortune in money to construct a workable apparatus of this kind, but it would not be impossible.

When, however, we tackle the question of the recognition by the automaton of a number of similar but not identical forms, the only solution we can think of is one which simply enlarges the repertory of the automaton by a number of *definite* shapes. It is, however, pretty certain that this is not Nature's way. The mental process is no doubt a sort of working to limit gauge. The same shape is recognized as the same until distortion has proceeded in various directions up to a certain point. But in the mental process, the shape appears in some way to be taken out of the three dimensions of space. The mirror image of a completely unsymmetrical shape is at once recognized as the same, or nearly the same, as the original.

Colour. — Colour is an entirely different matter. The recognition of colour by a sensitive relay would be quite a simple matter, for light can be filtered by screens so as to pick out any desired part of the spectrum, or, if that be not fine enough, a prism may be used. The applications to industry will be numerous. Wherever the human hand is guided by colour, the automaton could take its place. Thus in hand-picking and sorting by colour of natural objects, such as minerals, fruits, eggs, and so on, automata can be safely predicted. Mention has already been made in the press of an automatic device for picking out cigars of bad colour. Chemical processes are often controlled by observations of colour, turbidity, or other optical properties of the substances, and all these controls could well be made automatic.

Colour as a Label for Automata. — We can readily imagine that colour might be chosen as a convenient label to attach to objects of different sorts which are to be dealt with automatically. Cards could readily be sorted in accordance with their colours, and this might find application in automatic systems of accounting and book-keeping, instead of, as at present, the use of a code of holes punched in the cards. A standard-sized envelope with a certain repertory of colour schemes might be used to enable the automatic sorting of letters. One imagines that the labour-saving despatch of letters in the future will involve automatic correct franking according to weight, by the insertion of a coin, or, in the case of business houses, by the use of a franking counter of the kind recently introduced. The letters would then be automatically sorted by colour, it being, of course, a matter for the sender to use an envelope of the correct

125

colour, appropriate to the destination. The old-fashioned style of things could be kept up alongside the new, with appropriate excess of cost to those who preferred to send their letters in envelopes of their own choice.

It is obvious that the light-relay could equally well be made to respond to patterns printed in black on the white envelope, each destination having a particular key pattern.

Even at present, it is surprising that the post-offices of the world do not combine to bring pressure to bear in the direction of uniformity, of stationery, address, and so on. Hand-sorting would be greatly facilitated. . . .

The Mass Production of Machines. — We may now consider a field of work not without its analogies to the one last considered, namely, the mass-production of machines and contrivances, from motor cars to small matters such as vacuum cleaners. Here the mechanical production of the separate parts is brought to the highest stage of perfection by automatic tools working on material of great uniformity. The fundamental principle is that of working to limits of size. A very careful study of the machine to be mass-produced renders it possible to assign to every part certain limits of size within which that part may vary without losing the property of being interchangeable. It is, of course, impossible for the finest machine to produce even two objects exactly alike in every respect, let alone a large number. The tools used for stamping or punching or turning are subject to wear. The machines used never run perfectly truly. The material also varies in size and hardness to some extent. Changes in temperature, speed of machine, and lubrication all result in some change, no matter how minute, in the product. But we know by experience what it costs to produce objects in large numbers with any *stated* degree of accuracy as to size, that is to say, so that all the articles produced shall be smaller than one and larger than the other, of two given limiting sizes. The greater the accuracy, the greater the cost. Hence it is possible to calculate quite accurately the cost of mass-producing a given machine when the limits for each part have been calculated.

The object of production of parts to limits is to abolish "fitting" in the process of assembling the parts to form the finished machine. The parts are to go together just as if they belonged to a working machine which had been taken apart and was being put together again. Assembling thus becomes a purely mechanical operation, the more so as it is usual to inspect the finished parts before handing them over to be assembled. They are tested by what are called limit-gauges. Thus the limit-gauge for the boring of a cylinder consists of two plugs of metal, one slightly larger than the other. If the large one refuses to enter the cylinder, while the small one can be inserted, then we know that the size of the cylinder is intermediate between the two plugs.

Automatic Testing and Assembling. — How far is it likely that these two operations, of assembling and testing, will be performed by automata? As regards testing, there is a continual progress in the direction of rendering it automatic. It calls for labour of a responsible kind.

But assembling is quite another matter. Modern motor factory methods are sufficiently familiar to need no description here. Assembling itself is broken up into a large number of separate simple stages, while the rate at which the worker at each stage works is fixed by the rate at which the work comes to him to be done. He develops in the highest degree pure unthinking skill at the particular operation. To replace him by a machine would, no doubt, be perfectly possible in almost all cases, but only by designing a machine especially to perform that particular operation. But the capital cost and amorticisation although they might well still show a profit on hand labour, would add heavily to the already very heavy capital investment. After all, no one can say for certain that any such invention will be profitable, since sudden changes of fashion or technical methods may well render the apparatus out of date and unsaleable, while still in process of production.

There is also another point to consider. Automatic assembling would be greatly facilitated by the use of entirely different methods of construction. For instance, nuts and bolts are designed for assembly by human hands. A machine to insert a bolt and screw home nut and lock-nut would be a fairly complicated affair, but it would probably be possible to replace the nut and bolt by another type of fastening more amenable to the automaton. But the finished machine is destined for use in circumstances which may make repair necessary, or replacement of worn parts, and this has to be done by hand. Hence the machine must be constructed so that it can be readily taken apart by hand. . . .

Counting. — We may now consider another case in which the call upon the human intellect is of the simplest description, and hence the field for automata is a promising one, namely, the operation of counting. The application of automatic methods is here making rapid progress, but there are many directions in which applications of mechanical counting could still be made. For instance, the annual disturbance of stock-taking might be largely avoided. Very simple and cheap counters could be made to record, on a central board, the taking in and taking out of stock as it proceeds. The counter on the board would indicate the stock held at the moment. Instead of this, work is often at a standstill while time is spent counting thousands of articles of trifling value, for the purpose of the insertion of that value into a balance sheet. As, however, the other items on the balance sheet are estimated, very often, with a margin of accuracy of hundreds or thousands of pounds, the estimation of the stock to the nearest penny is an unscientific procedure.

Only quite recently was the proposal to introduce electrical indicators of the number and position of seats vacant in a theatre or cinema carried into practice in America. The system could readily be extended. One might be enabled to select a vacant seat and pay for it by the insertion of a coin in a machine, receiving a check which would serve as a key to enable that seat to be tipped down for use.

Accounting. — There is no need to remind the reader of all that has been said concerning the life at the ledger. Until the development of mass production, it stood uniquely for all that is most monotonous, mechanical, and dreary in human breadwinning.

Machinery is already in use by which a very great deal of the purely mechanical work can be eliminated. It depends fundamentally upon the same principle as the well-known Jacquard loom. Whatever is to be recorded is translated into a code of holes punched in a card. This card is then inserted into a machine, which is supplied with fingers or electric contacts which come up and look, as it were, for the holes. According to the combination of holes found, certain records are made, or the card may be passed on to another machine to be dealt with. We may imagine an order coming in for so many size so and so things at so much to be delivered in such and such a time. All this may be translated into the hole code by a single clerk. The subsequent operations of booking the order, notifying the correct department, checking the delivery time, totalling by classes all orders received, deducting goods when taken from stock, and invoicing them, could be performed mechanically. Moreover, the hole code can always be made to translate itself back to legible type, though the reverse operation would fall under one of those very difficult cases which we dealt with earlier. . . .

Light. — Another hopeful line of development consists in the possibility of finding some means of physical connection with earth other than light rays. Other kinds of ray are hardly practicable, though infra-red

The Shape of Things to Come: Glimpses of the year 2000 by a Gallic visionary of the year 1900. (From the private collection of Asa Briggs. Photographs by Michael Holford)

rays have some dim chance of utilization. The difficulty is, that we must needs form an image in an optical system of objects on the earth. Obviously, when there are no objects, even seeing the earth is no use to us, as over the sea. The use of the earth's magnetic field has been suggested. A wire stretched across the plane at right-angles to its direction of motion is cutting the magnetic lines of the earth's field at a rate proportional to the true speed of the plane. Hence, according to the principles of electromagnetism, a minute electrical voltage is developed between the ends of the wire, just as if it were a wire of the armature of a dynamo. We might hope to measure this if we could lead it back to our voltmeter. Unfortunately, the wires we should need to use would also cut the earth's lines, and hence neutralize the effect on the instrument. This is an interesting problem for those who know something of electricity and magnetism to puzzle over. If it could be solved, we should at once be able to tell both the true direction and the true speed of the plane, if we knew the strength and direction of the earth's field at the place of observation. We should thus have to know the position of the plane pretty accurately.

Another possible plan is as follows. If you move a magnet near a conductor, the magnet experiences a slight drag, due to the eddy currents induced by its field in the conductor. The earth being a conductor, a magnet on the plane experiences this drag. Unfortunately, a magnet is a heavy thing compared with the drag it experiences, and the earth is a bad, and worse still, a variable conductor. The drag on any conceivable magnet at any reasonable height would be excessively minute. But of recent years we have grown accustomed to measuring the incredibly minute.

The Railways. — Let us pass from considering the most recent means of transport to the question of one which is now regarded by many as doomed to gradual extinction; I mean the railway. One may truly say that the running of most of our railways is almost medieval from the standpoint of modern technical possibilities. Miniature railways are run in many towns for postal purposes entirely automatically. Very nearly the same thing could be done with a full-sized railway transporting passengers and goods, provided of course, it was electrically driven. The tube railways have already shown the way. There is no reason at all why one should not purchase a ticket to Aberdeen at King's Cross in exactly the same way that one purchases a ticket at Liverpool Street for Dover Street, from the man at the barrier. There is no reason why the Scotch express should not be a train in charge of one man, with automatic doors. There is no reason at all why it should be greeted at intermediate stations by hosts of officials. There is no sense in the present elaborate system of dealing with luggage.

128

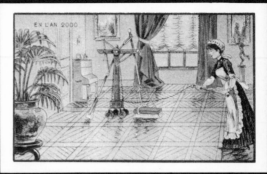

The taximan should dump it onto an automatic weighing machine which would discharge a printed ticket. A porter would then put it on the train in the same coach as the passenger, who would pay for it at the barrier as he took his ticket, on the base of the check from the weighing machine.

It is impossible to foresee whether the capital will be found to modernize our railways. . . . The driving could be made purely automatic; the word is a misnomer in connection with a really modern railway such as the Tube. The official in charge would press the starting button; the journey of the train to its stop at the next station would then be purely automatic, and safe against any mishap short of a landslide onto the line. If it arrived at a block in which another train still lingered, it would be stopped automatically. All this is possible by means already well tried. I am inclined to think that it will come sooner than we expect. . . . The capital cost of modernizing would be high, but the main item would be electrification, which is already being carried out rapidly on the Continent. The automata part of the business would be a comparatively small item of cost, provided that previous-experiment had enabled a standardized equipment to be produced by methods of mass-production. . . .

The Automaton in Power Generation. — One of the greatest advantages of the electrification of railways is the possibility of introducing the automaton principle in the generation of the power by which they are run. In few fields of modern technology is

it more widely applied than in power generation. One can buy quite cheaply small electric generating plants which are complete automata; they are widely used in country places. In Denmark (but in no other country as far as I am aware) one meets almost everywhere small automatic plants driven by wind power. These all depend upon the principle of keeping a battery of accumulators charged. As soon as an instrument indicates a drop in the charge, the dynamo is automatically set in motion to remedy the defect.

In large power stations such a use of accumulators would be out of the question, but the automatic regulation of the voltage produced by the dynamos has reached a state of the highest perfection. The periodicity, in the case of alternating current, is also regulated with the greatest exactness, and a by-product of this accuracy will in the future be of great importance in connection with automata.

A New Kind of Clock. — Modern current supply is almost exclusively alternating, that is, the current changes its direction many times (usually 50) per second. Now alternating current may be used to drive motors of many different types, one of which is the so-called synchronous motor. This type of motor works by exactly keeping time in its rotation with the alternations of the current; the number of revolutions it makes thus bears some simple proportion to the number of alternations of the current. It has either to work at this speed and in this way, or not at all. Hence, if the rate of the cur-

129

rent-alternations is timed very exactly, the motor which it drives also revolves with the same constant accuracy of speed. Such a motor may then be used to turn a clock train in place of spring and pendulum or balance. So accurate is the alternation of modern current, that a clock run in this way keeps good time. Furthermore, it is only necessary to have a clock of this kind at the central station, and observe its timing, in order to make any slight correction in the rate of alternation of the current. All other clocks connected to the network will then be corrected at the same time. The immense advantage of such a clock, apart from its simplicity, is the unlimited power behind it. Its rate does not depend upon the work it is called upon to do. So long as this is not great enough to stop the motor altogether, the clock runs true to time.

Inventors and Clocks. — From the earliest times inventors have had visions of doing marvellous things automatically by the use of clocks. They have always been hampered by the fact that you cannot get a clock to go accurately if you try and make it do work. Hence, it is necessary to use clocks so large and powerful that the work they are called upon to do is negligible in comparison with the power of the clock. All this trouble can now be banished. It is possible to make a simple and robust mechanism which will do anything in reason at any time you like to set it for.

The Time-Switch. — Before long, it should be possible to buy at a low price a clock which will switch on or off such currents as are used in everyday work at any predetermined time. Such "Time-switches" have, of course, been on the market for very many years, and have been widely used for controlling the lighting of streets, and for other purposes. On the whole, however, they have been found to be more bother than they are worth, for to be good they had to be fairly expensive. The synchronous-motor switch could be made, in large numbers of course, for a few shillings.

Automata in the Home. — We are thus led to the question of the place automata will occupy in the home of the future, assuming that such a thing as a home in the old-fashioned sense will exist at all. A few years ago, it was a favourite pastime of rich Americans to construct automatic homes. The door opens automatically when a member of the family, but not a stranger, approaches. The door-mat automatically brushes your boots. The thermostatic oven has been switched on at the correct time in anticipation of your arrival, and a touch of a switch causes it to discharge its contents onto a table, which then travels to the dining-room. When the dinner is eaten, another switch dismisses the remains of the feast to the kitchen. The romance of science used to go on like this for pages, but I suspect that the staff of skilled mechanics and electricians necessary to keep the whole business going proved more expensive than normal service, even in America.

Nevertheless, the time-switch should find many uses in the home of the future, especially in conjunction with electric cooking and heating. It has already been found capable of heating water and making tea, and waking the sleeper when these are ready for him. Whether it would be necessary to have it go to the length of turning the bed upside down, would depend upon the sleeper.

Answering the Telephone Automatically. — Another great need for the modern home is an automatic apparatus to take telephone messages. It would be somewhat costly, but would certainly find a large number of users, if satisfactory. It is one of those inventions which has been announced again and again, but it is to be supposed that a workable apparatus has not yet been produced, or at any rate, not put on the market. So many new developments which would seem to facilitate it have been made recently, that one may expect that a new effort will be made to achieve success. It is to be hoped that the introduction of automatic telephones will not render the application of a

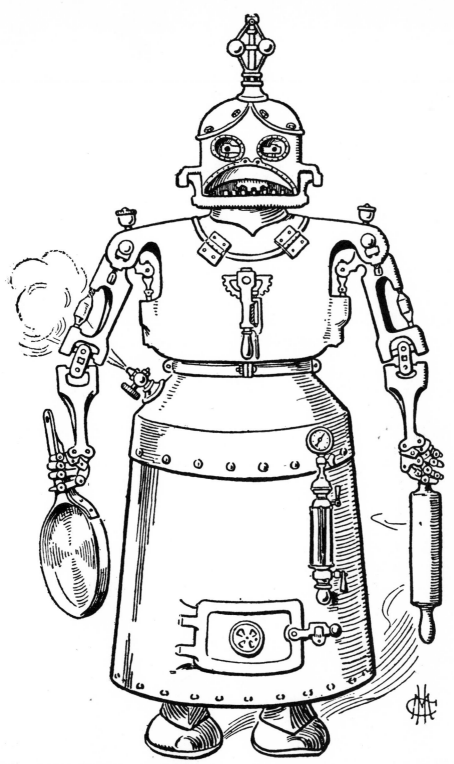

Contemplating this horrendous 1906 vision of "In 2000 — The Cook of the Future" from Puck, one wonders what even more appalling vision the artist must have had of the future housewife. (Historical Pictures Service, Chicago)

The figure contains the following labels: Visites MONSIEUR, MADAME, FOURNISSEURS, AEROCAB, Mademoiselle Monsieur, Madame, Cie G^le ALIMENTATION, Dejouber Diner, EXTRAS, Avance, Retard, Arret, Service, Blanchisserie, Chaussures, Barbe, Eau, Eclair, Trotteur, Bain

The maidservant of the future — another vision from the beginning of the present century — is conceived by artist Albert Robida to be a complex machine totally lacking in any anthropomorphic associations — let alone sex appeal! (New York Public Library, Picture Collection)

recorder any more difficult. One can imagine a device which would automatically raise the receiver from the hook when the bell had rung in vain a certain number of times, and apply it to the sound-recording device. The latter might well be photographic or magnetic, rather than the usual wax cylinder. A certain signal would be given to the caller, and he would then have a certain time in which to give his message; a most salutary arrangement. At the end of this time the receiver would sink back onto the hook. The sound recorder might be serviceable as a dictaphone for office use as well.

The Control of Domestic Heating. — Thermostatic control of domestic heating is long overdue. Here again the time switch could be used to cut the heating in and out at predetermined times. This would be easiest, of course, with electric heating, but would not present any great difficulty in connection with steam or gas, if any electric supply was available to work the necessary valves.

The satisfaction of going to bed, knowing that one would be called at a precise moment to a hot bath, tea, and a warm room, would be very great. Perhaps we might feel a little doubtful about the return in the evening, for it would be annoying, if unexpectedly detained, to know that all the automatic preparations for one's return were taking place with the inevitability of Fate. . . .

The Humanitarian's Opportunity. — It is a curious fact that the humanitarian has hitherto done little or nothing to do away with the dreary, monotonous, and disgusting labour inseparable from industry as carried on to-day. Instead, as we saw at the outset, his efforts to alleviate the lot of the masses actually compete against the efforts of science and invention to abolish manual labour. The supersession of hand-labour is decided almost entirely on financial grounds by the management of industry, and the certainty that cheap and contented labour will always be available, calling for no capital investment, tells heavily against the automaton. If the humanitarian really feels that the type of population resulting from his best efforts is a worthy and desirable product, there is nothing one can oppose to him that has not been said by the advocates of Mediævalism, craftsmanship, and the peasant life. But if there are humanitarians who have their doubts, one might suggest that they should seriously consider the employment of their resources on behalf of, and not against, the automaton.

Hitherto, comparatively trifling prizes have been offered, from humanitarian motives, for inventions designed to abolish particularly poisonous or dangerous occupations conducted by hand-labour. If, however, the whole weight of humanitarian influence were directed towards the substitution of automata for all forms of monotonous hand-labour, great progress might be made. The lot of the workers in the field should be made as attractive as that of the pure research worker. The idea, inevitable at a time when the problem of employment is so serious, that the first function of industry is to find employment for the masses, must be abolished in favour of the endeavour to make it supply our needs with the minimum of labour.

The supreme desire of our civilization must be realized, if its old age is to be honourable. The masses of human automata of which it makes use to-day must not be found among our descendants. Not the manual labour of teeming populations, but the power of great automata, must supply its physical needs; and, over and above these, resources for yet undreamed-of efforts to conquer matter, space, and time.

Karel Čapek's play, R.U.R.
(Rossum's Universal Robots), A Fantastic
Melodrama ..., 1922: (Theatre Collection,
Library and Museum of the Performing Arts,
New York Public Library at Lincoln Center)

MAN
VERSUS
MACHINE?

Through actual robots as they are conceived, created and utilized in our society, man vividly expresses myriad aspects of himself — his humor, playfulness, ingenuity, industry, aggression, self-indulgence. The illustrations in this section show some of the ways in which the machine has become a measure of modern man.

Darwin Among the Machines

by Samuel Butler

The following letter was first published in *The Press*, Christ Church, New Zealand, June 13, 1863. It was reprinted by H. Festing Jones in his edition of *The Notebooks of Samuel Butler* (1912) together with an editor's note observing that the letter was Butler's earliest expression of ideas about man and machine that were to be developed in the novel *Erewhon* (1872).

The serious implications of the letter — particularly its closing paragraphs — have increasingly overshadowed the delicious irony that suffuses Butler's writing. This remarkable *jeu d'esprit* by the young Samuel Butler (he wrote it when he was twenty-eight) almost casually anticipates some of the direst fears and warnings of the anti-utopian pundits, prophets, and science fiction writers of the twentieth century. In the brief period of calm before he was to pour out his wrath against Darwin and Darwinian thought, Butler playfully drew analogies between the development of machines and the evolutionary process as it was conceived in *The Origin of Species* (1859). Shortly after writing "Darwin Among the Machines," Butler was to launch his lonely, lifelong attack against Darwinism. As Bernard Shaw noticed in the Preface to *Back to Methuselah* (1921), Butler, realizing that Darwin conceived evolution as a purposeless process, "declared with penetrating accuracy that Darwin had 'banished mind from the universe'; and even attacked Darwin's personal character, unable to bear the fact that the author of so abhorrent a doctrine was an amiable and upright man. Nobody would listen to him. He was . . . completely submerged by the flowing tide of Darwinism." Nevertheless, in a series of works such as *Life and Habit* (1877), *Evolution, Old and New* (1879), *Unconscious Memory* (1880), and *Luck or Cunning?* (1887), he developed his philosophy of Vitalism or Creative Evolution in opposition to Darwinian thought.

Sir — There are few things of which the present generation is more justly proud than of the wonderful improvements which are daily taking place in all sorts of mechanical appliances. And indeed it is matter for great congratulation on many grounds. It is unnecessary to mention these here, for they are sufficiently obvious; our present business lies with considerations which may somewhat tend to humble our pride and to make us think seriously of the future prospects of the human race. If we revert to the earliest primordial types of mechanical life, to the

From H. Festing Jones, ed., *The Notebooks of Samuel Butler* (London: Fifield, 1912).

137

lever, the wedge, the inclined plane, the screw and the pulley, or (for analogy would lead us one step further) to that one primordial type from which all the mechanical kingdom has been developed, we mean to the lever itself, and if we then examine the machinery of the *Great Eastern*, we find ourselves almost awestruck at the vast development of the mechanical world, at the gigantic strides with which it has advanced in comparison with the slow progress of the animal and vegetable kingdom. We shall find it impossible to refrain from asking ourselves what the end of this mighty movement is to be. In what direction is it tending? What will be its upshot? To give a few imperfect hints towards a solution of these questions is the object of the present letter.

We have used the words "mechanical life," "the mechanical kingdom," "the mechanical world" and so forth, and we have done so advisedly, for as the vegetable kingdom was slowly developed from the mineral, and as in like manner the animal supervened upon the vegetable, so now in these last few ages an entirely new kingdom has sprung up, of which we as yet have only seen what will one day be considered the antediluvian prototypes of the race.

We regret deeply that our knowledge both of natural history and of machinery is too small to enable us to undertake the gigantic task of classifying machines into the genera and sub-genera, species, varieties and subvarieties, and so forth, of tracing the connecting links between machines of widely different characters, of pointing out how subservience to the use of man has played that part among machines which natural selection has performed in the animal and vegetable kingdoms, of pointing out rudimentary organs[1] which exist in some few

machines, feebly developed and perfectly useless, yet serving to mark descent from some ancestral type which has either perished or been modified into some new phase of mechanical existence. We can only point out this field for investigation; it must be followed by others whose education and talents have been of a much higher order than any which we can lay claim to.

Some few hints we have determined to venture upon, though we do so with the profoundest diffidence. Firstly, we would remark that as some of the lowest of the vertebrata attained a far greater size than has descended to their more highly organised living representatives, so a diminution in the size of machines has often attended their development and progress. Take the watch for instance. Examine the beautiful structure of the little animal, watch the intelligent play of the minute members which compose it; yet this little creature is but a development of the cumbrous clocks of the thirteenth century — it is no deterioration from them. The day may come when clocks, which certainly at the present day are not diminishing in bulk, may be entirely superseded by the universal use of watches, in which case clocks will become extinct like the earlier saurians, while the watch (whose tendency has for some years been rather to decrease in size than the contrary) will remain the only existing type of an extinct race.

The views of machinery which we are

---

[1] We were asked by a learned brother philosopher who saw this article in MS. what we meant by alluding to rudimentary organs in machines. Could we, he asked, give any example of such organs? We pointed to the little protuberance at the bottom of the bowl of our tobacco pipe. This organ was originally designed for the same purpose as the rim at the bottom of a tea-cup, which is but another form of the same function. Its purpose was to keep the heat of the pipe from marking the table on which it rested. Originally, as we have seen in very early tobacco pipes, this protuberance was of a very different shape to what it is now. It was broad at the bottom and flat, so that while the pipe was being smoked the bowl might rest upon the table. Use and disuse have here come into play and served to reduce the function to its present rudimentary condition. That these rudimentary organs are rarer in machinery than in animal life is owing to the more prompt action of the human selection as compared with the slower but even surer operation of natural selection. Man may make mistakes; in the long run nature never does so. We have only given an imperfect example, but the intelligent reader will supply himself with illustrations.

Rickshaw Robot. In the year 1868, citizens of Newark, N.J., were given an interesting glimpse of a combination of Chinese servility and American ingenuity when this mechanical rickshaw man built by Zadoc P. Dederick took to the streets. (Brown Brothers)

ourselves the inferior race. Inferior in power, inferior in that moral quality of self-control, we shall look up to them as the acme of all that the best and wisest man can ever dare to aim at. No evil passions, no jealousy, no avarice, no impure desires will disturb the serene might of those glorious creatures. Sin, shame, and sorrow will have no place among them. Their minds will be in a state of perpetual calm, the contentment of a spirit that knows no wants, is disturbed by no regrets. Ambition will never torture them. Ingratitude will never cause them the uneasiness of a moment. The guilty conscience, the hope deferred, the pains of exile, the insolence of office, and the spurns that patient merit of the unworthy takes — these will be entirely unknown to them. If they want "feeding" (by the use of which very word we betray our recognition of them as living organism) they will be attended by patient slaves whose business and interest it will be to see that they shall want for nothing. If they are out of order they will be promptly attended to by physicians who are thoroughly acquainted with their constitutions; if they die, for even these glorious animals will not be exempt from that necessary and universal consummation, they will immediately enter into a new phase of existence, for what machine dies entirely in every part at one and the same instant?

We take it that when the state of things shall have arrived which we have been above attempting to describe, man will have become to the machine what the horse and the dog are to man. He will continue to exist, nay even to improve, and will be probably better off in his state of domestication under the beneficent rule of the machines than he is in his present wild state. We treat our horses, dogs, cattle, and sheep, on the whole, with great kindness; we give them whatever experience teaches us to be best for them, and there can be no doubt that our use of meat has added to the happiness of the lower animals far more than it has detracted from it; in like manner it is

thus feebly indicating will suggest the solution of one of the greatest and most mysterious questions of the day. We refer to the question: What sort of creature man's next successor in the supremacy of the earth is likely to be. We have often heard this debated; but it appears to us that we are ourselves creating our own successors; we are daily adding to the beauty and delicacy of their physical organisation; we are daily giving them greater power and supplying by all sorts of ingenious contrivances that self-regulating, self-acting power which will be to them what intellect has been to the human race. In the course of ages we shall find

reasonable to suppose that the machines will treat us kindly, for their existence is as dependent upon ours as ours is upon the lower animals. They cannot kill us and eat us as we do sheep; they will not only require our services in the parturition of their young (which branch of their economy will remain always in our hands), but also in feeding them, in setting them right when they are sick, and burying their dead or working up their corpses into new machines. It is obvious that if all the animals in Great Britain save man alone were to die, and if at the same time all intercourse with foreign countries were by some sudden catastrophe to be rendered perfectly impossible, it is obvious that under such circumstances the loss of human life would be something fearful to contemplate — in like manner were mankind to cease, the machines would be as badly off or even worse. The fact is that our interests are inseparable from theirs, and theirs from ours. Each race is dependent upon the other for innumerable benefits, and, until the reproductive organs of the machines have been developed in a manner which we are hardly yet able to conceive, they are entirely dependent upon man for even the continuance of their species. It is true that these organs may be ultimately developed, inasmuch as man's interest lies in that direction; there is nothing which our infatuated race would desire more than to see a fertile union between two steam engines; it is true that machinery is even at this present time employed in begetting machinery, in becoming the parent of machines often after its own kind, but the days of flirtation, courtship, and matrimony appear to be very remote,

and indeed can hardly be realised by our feeble and imperfect imagination.

Day by day, however, the machines are gaining ground upon us; day by day we are becoming more subservient to them; more men are daily bound down as slaves to tend them, more men are daily devoting the energies of their whole lives to the development of mechanical life. The upshot is simply a question of time, but that the time will come when the machines will hold the real supremacy over the world and its inhabitants is what no person of a truly philosophic mind can for a moment question.

Our opinion is that war to the death should be instantly proclaimed against them. Every machine of every sort should be destroyed by the well-wisher of his species. Let there be no exceptions made, no quarter shown; let us at once go back to the primeval condition of the race. If it be urged that this is impossible under the present condition of human affairs, this at once proves that the mischief is already done, that our servitude has commenced in good earnest, that we have raised a race of beings whom it is beyond our power to destroy, and that we are not only enslaved but are absolutely acquiescent in our bondage.

For the present we shall leave this subject, which we present gratis to the members of the Philosophical Society. Should they consent to avail themselves of the vast field which we have pointed out, we shall endeavour to labour in it ourselves at some future and indefinite period.

I am, Sir, etc.,

CELLARIUS

Is Man a Robot?

by John Cohen

In an article in Sidney Hook's symposium, *Dimensions of Mind* (New York: New York University Press, 1960), Professor Michael Scriven discusses what he considers to be the basic activities that a synthetic being would have to be capable of in order to qualify as a "person:" i.e., motion, reproduction, perceiving, feeling, learning, understanding, interpreting, analyzing, making discoveries through trial and error, choosing between alternatives, decision-making, creating, and lying. Scriven concludes: "A robot might do many of . . . [these] things . . . and not qualify [as a human]. It could not do them all and be denied the accolade." By contrast, Professor Cohen maintains that certain human characteristics — laughter and tears, blushing, and suicide — will remain beyond the capabilities of robots. An embarrassed robot who bursts into tears and then hurls itself off a cliff would thus have to be regarded as a human being. But the construction of such a sensitive robot is beyond the present and foreseeable potentialities of man.

I

Is man a mindless robot? or can he boast an unrobotlike mind? The meaningfulness of these questions hangs on what we mean by 'mind', and this we shall soon try to make clear. Throughout the foregoing chapters these questions have haunted us as well as the automaton-makers, and we cannot decently take leave of our theme before confronting them, even if, like St. Augustine, we have to reply 'I know not'.

We need not take the dogmatic position that a man's ideas are wholly shaped by his mother tongue. Nor would it be wise to assume that usage of the English word 'mind' happens to be the sole repository of truth about a supremely baffling problem of philosophy and psychology. If we took as our point of departure a word or phrase which had little or nothing to do with Western tradition we might arrive at totally different conclusions. We could begin with the Chinese *Hsin li* ('the reasons or principles of the heart');[1] with the Egyptian *ka, ba, akh,* or *khaba* (shade, soul, mind, image); with the Hebrew *nephesh, ruah,* or *neshamah* (life, spirit, soul); with the Greek *nous, logos,* or *pneuma* (mind or spirit, wisdom, breath); with the Latin *anima* or *animus* (breath, rational soul); with the Japanese *kokoro, tamashi, seishin* (mind, soul, spirit) or with any of the countless expressions for 'mental' functions in archaic or surviving primitive

From John Cohen, *Human Robots in Myth and Science* (London: George Allen & Unwin, 1966). Reprinted by permission.

[1] I. Veith, Non-Western Concepts of Mental Function, pp. 29–42 in *The History and Philosophy of Knowledge of the Brain and its Functions* (ed. F. N. L. Poynter), Oxford: Blackwell, 1958, p. 33.

languages. Let us avoid the trap of supposing that our problem will be solved if we confine ourselves to the English or any other particular language. At least one writer on 'The Concept of Mind' has been discomfited as a result of founding a theory of mental life on the grammatical structure of the English language and on 'the standards of correct English usage'.[2] Apart from the difficulty presented by words as such, it seems to be a fact that a particular conception of man and his nature is implicit in each of the multitude of 'world views' and ideologies. It has yet to be shown that any one of these conceptions is alone entitled to wear the badge 'scientifically approved'. The never-ending disputes on the subject in the Western world are enough to show that this is not the case.

The Indo-germanic root of the word 'mind' appears to be *men, mon-* to think, remember, intend. Sanskrit has *manas, man*, to think. The Anglo-Saxon *gemynd* mind, is cognate with Gothic *gamunds* memory, Latin *mens, memini* I remember, *moneo* I advise, and German *minne* love.[3] A full account even of these words would call for a history of psychology, which 'is nothing more nor less than a history of the different ways in which men have looked upon the mind'.[4] If we did limit ourselves to English usage, we should still have to recognize that this has been profoundly influenced by Christianity and earlier cults as well as by the gradual elaboration of ideas about mental life handed down

from Greek antiquity influenced by earlier Oriental cosmologies.

The complexity of the Greek influence will be appreciated if we recall that at least five stages may be distinguished in the development of the Greek idea of the psyche or soul from Homer to the philosophers of the fifth century BC: (i) a kind of breath which is blown out at death, (ii) the seat of emotion, (iii) intellectual 'interpreter' of sense data, (iv) a moral as well as an intellectual faculty, and (v) 'a person and within a person a soul' — the most precious part; the word *psyche* could at the same time denote a thing, a process or an agent (personal or divine).[5] Perhaps the two most important later Greek influences on English usage have been the dualism of Plato for whom the soul, as an immaterial entity quite distinct from the body, apprehends an ideal world, and the quasi-biological view of Aristotle for whom the soul or mind is process, form or function, defined in terms of its activity, and the body-mind is a natural and individual whole.

The gap of two thousand years between Aristotle and seventeenth century England is not hard to bridge. Among British thinkers, Hobbes and Locke were the first to adopt an empirical conception of mind. The passivity of mind, which marks Locke's view, remained a feature of British associationist psychology until John Stuart Mill introduced an active principle, 'mental chemistry', later to be elaborated by William James, Ward, Stout and MacDougall. On the European continent the conception of mind as an active agent may be traced to Leibnitz whose influence on Brentano and Külpe was specially notable. Brentano's original contribution to the distinguishing feature between bodily and mental phenomena is that the latter are invariably characterized by an intention. They are always directed towards

[2] J. R. Smythies, *The Analysis of Perception*, London: Routledge and Kegan Paul, 1956, p. 103.

[3] C. G. Jung draws attention to Lat. *mentula*, a poetic expression for the *membrum virile*, and adds that *mentula* might be a diminutive of *menta* or *mentha*, mint, which in antiquity was regarded by some as an aphrodisiac, or as erotically inhibiting, or as a contraceptive. He quotes a source to the effect that mint is called *menta* because its smell excites the mind, and he remarks that a strong or priapic chin was called *mentum*. (C. G. Jung, *Symbols of Transformation*, London: Routledge and Kegan Paul, 1956, pp. 146–7.)

[4] J. M. Baldwin, *History of Psychology*, London: Watts, 1913, p. 1.

[5] T. B. L. Webster, 'Communication of Thought in Ancient Greece,' pp. 125–146 in *Studies in Communication*, London: Secker and Warburg, 1955.

an object. Anger and love are not things like tables and chairs. We can only be angry with someone or about something. Love's target is likewise part of love. Thus, too, in all desire something is desired, in all judgements something is affirmed or denied, accepted or rejected, and all hatred encompasses a hated object. But Brentano still separates the subject from his world. He does not, like his existentialist successors, destroy this separation and treat the object as part of the 'intentionality' of the subject. The suggestion that all mental processes have an organized quality from the very start, that is, from birth, was introduced by the Gestalt theorists in their attack on the 'atomism' of the associationists, although the 'mind' of Gestalt theory is just as passive as that of the associationists.

Both the active and the organized qualities are present in the Freudian theory of mind as a dynamic structure of interrelated conscious and unconscious agencies, to which C. G. Jung added the 'archetypes' of the collective unconscious. Jung's 'archetypes' have received scant attention from academic psychology. It is all the more interesting therefore to see how they were interpreted by the late W. Pauli, who compared them to statistical laws of nature. Both conceptions, he wrote, share a 'tendency to amplify the older more narrow idea of "causality (determinism)" to a more general form of "connection" in nature, towards which the psycho-physical problem also points'.[6] Pauli refers, as well, to the similarity between ideas of 'correspondence', 'complementary pairs of opposites', and 'wholeness' in physics, on the one hand, and in theories of the unconscious, on the other.

An impressive attempt to trace the source and development of the formal properties of mind, as embodied in logico-mathematical reasoning, is made by Piaget in his genetic epistemology, a system which sets forth the supposed logical properties of reasoning at each stage in the child's mental growth, but it is not clear to what extent his explanations, as distinct from his observations, relate to psychological processes and to what extent, like Spearman's *noegenesis*, they constitute a logical scheme imposed on the observations. The issue is that raised by C. Wolff (1679–1754) as between 'empirical' and 'rational' psychology, a distinction which ultimately takes us back to the 'soul-biology' and 'soul-theology' of the Greeks.[7]

Many who find the word 'mind' useful normally understand by it a property of *individual* man, as Sherrington insisted, 'mind, always, as we know it, finite and individual, is individually insulated and devoid of direct liaison with other minds'.[8] But it is not unusual to come across references to a 'group mind' and to the 'minds' of animals. Such Western usage, in which 'mind' signifies a 'phenomenon' contrasts with religious psychology in the East where a connexion is assumed between individual 'mind' and a hypothetical Universal Mind.[9] The latter view is encountered in the writings of Western mystics, such as Meister Eckhart.

Naturally, there are not lacking theorists who hold that the word 'mind' is devoid of meaning: 'Psychology is supposed to be the study of mind, but since no one knows what is meant by mind, it is impossible to define psychology. . . . There is no such thing as mind — at least with a capital 'm' . . . for all scientific intents and purposes the concept has outlived its usefulness';[10] and C. A. Mace[11] declares that it is undesirable 'to

---

[6] W. Pauli, 'Naturwissenschaftliche und erkenntnis-theoretische Aspekte der Ideen von Unbewussten', Abstract in *Brit. J. Phil. Sci.*, 1956, 6, p. 259.

[7] M. Dessoir, *Outlines of the History of Psychology*, New York: Macmillan, 1912, p. 134.

[8] C. Sherrington, *Man on his Nature*, London: Macmillan, 1940, p. 261.

[9] C. G. Jung, *Psychology of Religion: West and East*, London: Routledge and Kegan Paul, 1958, pp. 475–6.

[10] C. C. Pratt, *The Logic of Modern Psychology*, New York: Macmillan, 1940, p. viii.

[11] Article 'Psychology', *Encycl. Brit.*, 1950.

embalm' the word 'mind' in any definition of psychology.

Broadly speaking, contemporary psychological opinion varies with the position adopted in relation to the traditional body-mind problem, and it ranges from idealism, which treats the world of psychical processes as subject to its own causal principles *sui generis*, to materialism for which 'mind' is an epiphenomenon, a ghost in the machine.

In Soviet Russia, some official spokesmen for psychology consider 'human consciousness as a subjective reflection of the objective world',[12] a formulation which bristles with difficulties: what is meant by 'reflection'?, by 'the objective world'? Apart from these semantic difficulties, the statement, in spite of its partial truth, seems at loggerheads with some facts, while leaving others completely unexplained.

II

I now turn to those aspects of the problem which loom large in current debates. A common source of error is the confusion between 'dualism' as a metaphysical interpretation of the body-mind relationship and 'dualism' as an actual description of experience without any implication as to the ontological status of mind and matter. 'Dualism' is not a serviceable word to use when we are called upon to do justice to experience by verbal description. There is rather a continuum between experiences which, at one extreme, unmistakably involve organic sensations, like hunger or sex, and at the other extreme, involve what seems subjectively to be pure, disembodied reflection. Intermediate between them are experiences, especially those of an affective character, which partake of both bodily and mental components to a greater or less extent. I speak of the normal range of experience. The phenomenon of *subjective* disembodiment is well-attested in psychopathology. We may thus, I think, justifiably consider 'the idea of mind' without becoming entangled in metaphysical controversy, though without denying that this also has the right to a logical existence.

Two main issues seem to arise in contemporary discussions of 'mind' and, indeed, by implication, in the domain of psychology as a whole. The first is the problem of reductionism.[13] We are told that 'both the formal and the final aspects of that activity which we are wont to call *mental* are rigorously deducible from present neurophysiology'.[14] It is hard to know what to make of this claim. Certainly little or nothing established in the experimental psychology of the higher mental process is deducible in this fashion. It is true that what everyone calls 'mental' activity is served by the brain. It is also true that a knowledge of neurophysiology is essential for a full understanding of sensory process, of memory, and of the effects of brain damage upon mental and behavioural functioning. But this is a far cry from deducing the processes of judgement, choice, decision-making, and belief from present-day neurophysiology. For similar reasons we are not impressed by the unsubstantiated claim that 'diseased mentality can be understood

12 V. S. Molodtsov, 'Philosophy, Sociology, Logic and Psychology', *Internat. Soc. Sci. J.*, 1959, 11, p. 183.

13 See F. M. R. Walshe, 'Current Ideas in Neurobiology and Neuropsychology: A Study in Contrasts', pp. 199–218 in *Perspectives in Biology and Medicine*, Vol. 7, No. 2, 1964, Chicago: University of Chicago Press; 'The Brain-stem conceived as the' 'highest level' 'of function in the Nervous System', *Brain*, 1957, 80, pp. 510–539; 'Thoughts upon the equation of Mind with Brain', *Brain*, 1953, 76, pp. 1–18.

14 Warren S. McCulloch and Walter H. Pitts, 'A Logical Calculus of the Ideas Immanent in Nervous Activity', pp. 379–99 in *Information Storage and Neural Control* (ed. W. S. Fields and W. Abbott), Springfield, Ill: Charles C. Thomas, 1963.

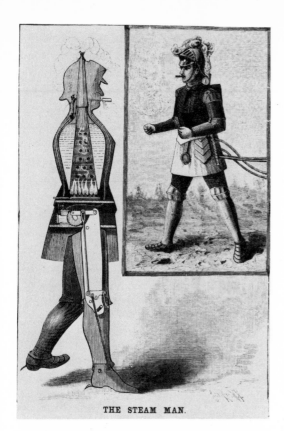

THE STEAM MAN.

This steam robot was built in Canada in 1893. Propelled by a half-horsepower motor which drove jointed rods to move the robot's legs, it could walk in circles at the end of a horizontal rod. Protruding from its mouth was an exhaust pipe from its gasoline-fired boiler. (The Bettmann Archive, Inc.)

without loss of scope or rigor, in the scientific terms of neurophysiology.'[15] Nor will a psychiatrist draw much comfort from the gratuitous dogma that 'for prognosis, history is never necessary'. This may be true for an automobile. It is not true for the driver.

There is one species of reductionism which takes the form of a demand for translatability. 'It is not necessary that the student of personality talks in neurological terms but his terms should be translatable when necessary into neurology.'[16] Why we should submit ourselves to this moral compulsion it is difficult to understand. Nor is it clear when

it *is* necessary and when it is *not*. If we surrendered to this self-denying ordinance our statements about personality would be limited to trivia. Certainly anything Freud wrote on the subject would have to be ignored. What is more, psychology would cease to exist, for it would no longer be permitted to make observations or construct theories which are psychologically distinctive *sui generis*.

Even so thoroughgoing a reductionist as Lashley[17] seems opposed to the test of translatability. 'Psychology', he declared, 'is today a more fundamental science than neurophysiology. By this I mean that the latter offers few principles from which we may predict or define the normal organization of behaviour, whereas the study of psychological processes furnishes a mass of factual material to which the laws of nervous action in behaviour must conform.' If we accept this statement, as I believe we should, the argument for translatability falls to pieces. This argument is associated with his curious hostility to the concept of 'mind'. Hebb insists that 'knowledge of the mind is chiefly theoretical, inferred from behaviour rather than being obtained directly from self-observation (i.e. from introspection)'.[18] If this means that our knowledge of mind is of inferior scientific status because it is chiefly theoretical, then the same low status must be ascribed to nuclear physics, because our knowledge of what is inside the atom is also theoretical. But we must reject the one-sided view that our knowledge of mind is inferred from behaviour. We might equally well argue that our understanding of behaviour depends on our capacity for self-observation. If we had no experience or awareness of love or fear no amount of observation of be-

[15] *Loc. cit.*

[16] D. O. Hebb, 'The Role of Neurological Ideas in Psychology', *J. Personal.*, 1951, 20, pp. 39–55.

[17] K. Lashley, *The Neuropsychology of Lashley*, New York: 1960, p. 207 quoted by Lord Brain, 'Body, Brain, Mind and Soul', pp. 51–63 in *The Humanist Frame* (ed. J. S. Huxley), London: Allen and Unwin, 1961. See also his 'Some Reflections on Brain and Mind', *Brain*, 1963, 86, pp. 381–402.

[18] D. O. Hebb, *Introduction to Psychology*, London: Saunders, 1950, p. vii.

145

haviour would enlighten us. It may be noted at this point that the contempt for introspection which Hebb enjoys has been abandoned by the positivists who gave it birth. 'The subject-matter of psychology', writes von Mises,[19] 'entirely within the framework of the empirical conception of science, can be characterized by the fact that in its structure enter element statements of a specific kind which may be summarized in the expression "self-observation".' After leading reductionists into the wilderness for thirty years Carnap[20] likewise recanted in words to the same effect.

This leads us to the second main issue. A basic flaw in reductionism and the neurological automatism which it entails is its impersonal character. Psychologically, though not neurologically, we can speak of a capacity for understanding meaning, which may be highly personal and sometimes incommunicable in words. Psychologically too, we can speak of belief, of powers of reflection and originality, and of 'reaching convictions by an act of personal judgement'. The reductionist, in asserting that the mental life of man can be *wholly* represented in terms of a neural automaton, denies to him those very qualities which distinguish him from a robot.[21]

III

A number of enterprising investigators are busily devising ever more complex automata intended to reproduce some of the essential functions of the human brain-mind. Grey Walter's mechanical 'tortoises' and CORA, and Ross Ashby's *homeostat* are among the most illustrious of these imitation men. Provided we do not assume that they embody all the essential properties of man, they may serve to illumine the nature of human abilities and disabilities. In spite of his critics, Grey Walter makes no exaggerated claims on behalf of his own automata. They bear about as much relation to the human brain, he says, as a hacksaw does to the human hand.[22]

In his brilliant essay on Computing Machinery and Intelligence, the late A. M. Turing considered the theological objection to the view that machines can think. A theologian might argue that God has given a thinking soul only to man but not to any animal. But, says Turing, it would not be beyond the power of an ominipotent Al-

mighty to confer an immortal soul on an elephant, even if He felt it would only be appropriate to do so if the elephant were to be equipped with a brain capable of ministering to this soul. The idea of a thinking machine is no more sacrilegious than the idea of an elephant with a soul, and if we made a machine that thinks (says Turing) 'we should not be irreverently usurping His power of creating souls, any more than we are in the procreation of children: rather we are, in either case, instruments of His will providing mansions for the souls that He creates'.

It would seem that at least three things characteristically human are out of reach of contemporary automata. In the first place, they are incapable of laughter (or tears); secondly, they do not blush; thirdly, they do not commit suicide. It is conceivable that robots of the future may be capable of all three. However, until we have a better understanding of the nature of laughter it would be unwise to assume that we shall be

---

19 R. von Mises, *Positivism*, Cambridge, Mass.: Harvard University Press, 1951.

20 R. Carnap in *Minnesota Studies on the Philosophy of Science* (ed. H. Feigl [and] M. Scriven), Minneapolis: University of Minnesota Press, 1956.

21 M. Polanyi, *Personal Knowledge*, London: Routledge and Kegan Paul, 1958, 263.

22 W. Grey Walter in *Discussions on Child Development*, II, (ed. by J. M. Tanner and B. Inhelder), London: Tavistock Publications, 1956.

able to teach robots how to laugh. The problem is rendered more complex by the fact that there seems to be a double relationship between the laughable and the automatic. We laugh when we see a human being behave like an automaton, if a speaker, for example, moves his head in a tic-like, stereotyped fashion. The appearance of mechanism where we expect life provokes laughter. And conversely we laugh when a true robot behaves like a man, and the closer the resemblance the more comical we find the situation. It is a little hard to imagine a true robot laughing because another true robot seems lifelike or, alternatively, because its designer appears to conduct himself like a fellow robot.

Blushing may not turn out to be much more manageable although both the anatomical and the psychological processes involved in blushing are fairly clear. The reader may wonder why blushing rather than other features of man is highlighted here. The answer is that blushing seems a singularly human phenomenon. It belongs to the expressive language of the human face. Its anatomical basis is an intricate system of capillaries which line the inner walls of the cheeks and which have a network of nerve fibres to serve them. This capillary action is the means of making our private feelings visible to an observer, for they make the blush possible. Monkeys flush in anger but they cannot be said to blush in shame. Perhaps the transition from flush to blush constitutes the dividing line between man and animal. Darwin[23] called blushing 'the most wondrous of all the wondrous powers of the mind . . . and the most human of all expressions'. We blush when we feel exposed, physically or mentally, when we have been unmasked, when we have made what others see is a stupid mistake, when caught red-handed, when wrongly accused. We blush

when we merely think about what someone else is thinking of what we are thinking. The common factor in these various situations is that we feel ourselves caught in an impasse. We cannot for the moment find a way out; so there is nothing to do but to blush. The blush is an outward manifestation of what subjectively is experienced as shame in the presence of others, and it takes place in the face because our experience of being in the presence of others is somehow localized in the face, which is that part of us offered for public display. But we do not only blush for shame. Darwin remarked that a pretty girl blushes when a boy gazes at her even though she knows perfectly well that his gaze is one of admiration. Her blush draws attention to herself while enhancing her charms.

As Buytendijk[24] has remarked, with his customary shrewdness, girls blush more than boys because the significance of exposure is not the same for them as it is for boys. And this is due to the fact that a girl's relationship to her body differs from a boy's relationship to his, just as her relationships to other people are different from his. The adolescent girl, unlike her brother, feels her relationship to others mediated through her body, and particularly through her clothes, which serve the ambiguous purpose of covering and revealing at one and the same time. Adolescence, furthermore, is a phase when girls are more sensitive than ever to their appearance. The merest glance can provoke a blush, and the girl feels helpless, as if her protective covering had been torn aside.

In the light of this, we cannot now foresee how a future computer could be programmed to blush in suitably embarrassing circumstances, and we have to bear in mind that it is just as human not to blush when we should as to blush when we shouldn't.

Thirdly, suicide on the part of any future

---

[23] C. Darwin, *The Expression of the Emotions in Man and Animals*, London: 1872.

[24] F. J. J. Buytendijk, The Phenomenological Approach to the Problem of Feelings and Emotions' in *Feelings and Emotions* (ed. M. L. Reymert), New York: McGraw-Hill, 1950, pp. 127–141.

The Epilogue of Karel Čapek's play, R.U.R.
(Rossum's Universal Robots), A Fantastic
Melodrama . . . , *1922:*

> RADIUS. Master, the machines will not do the
> work. We cannot manufacture Robots.
> FIRST ROBOT. We have striven with all our
> might. . . .
> SECOND ROBOT. Eight million Robots have died
> this year. Within twenty years none will be left.
> FIRST ROBOT. Tell us the secret of life. . . . Teach
> us to multiply or we perish. . . . We cannot beget
> children. Therefore, teach us how to make
> Robots. . . .
> ALQUIST. I am the last human being, Robots, and
> I do not know what the others knew. . . . I
> cannot create life.

*Čapek's play gave the word "robot" to the
world, but the robots of his play are actually*
*androids who revolt against their creators and
eliminate all but one human being — the
builder Alquist. Ultimately, when Alquist is
about to dissect two "robots" — Helena and
Primus — to discover the secret of their manufacture, they rebel against him, prepared to
destroy themselves rather than submit to him.
When Primus tells Alquist, "Man, you shall
kill neither of us. . . . We — we — belong to
each other," Alquist realizes that humanity is
not doomed after all. "Go. Adam — Eve," he
says to Primus and Helena as they go forth to
repopulate the earth. (Theatre Collection,
Library and Museum of the Performing Arts,
New York Public Library at Lincoln Center)*

robot may have to be ruled out. A robot may be endowed with the capacity to bring about its own disorganization when conditions reach a given threshold of stress. But true suicide implies a foreknowledge of death and some idea of its significance,[25] and this is a privilege of man.

In general, whatever refinements and novelties are introduced into artifacts in the forseeable future, man is destined to remain for a very long time, the lightest, most reliable, most cheaply serviced and the most versatile general-purpose computing device made in large quantities by unskilled labour, an observation appropriately attributed to a naval officer.[26]

The logical system which is embodied in a computer is a *tool* which in and by itself is utterly useless. It requires for its logical completion someone who is able to use it in a fashion and for a purpose not fully predetermined by the tool. As Michael Polanyi[27] has made abundantly clear, to elevate the machine to the logical or psychological status of its maker is to commit the behaviourist's error when he confuses the observed 'mind' with the observing mind. The thought or purpose which we are able to detect in the working of a machine is properly speaking the thought or purpose of the designer of the machine, for the most cunningly devised automaton is, in principle, in the same logical class, in relation to its maker, as the most rudimentary flint scraper of primitive man.[28]

Moreover, however spectacular the achievements of future computers in solving intellectual problems, they will still leave much to be done in the more troublesome sphere of human relationships. We are therefore indebted to the perspicacity of Leopardi[29] (1798–1837) who conceived an agency which could control the moral as well as the material world. He wondered whether we should ever have 'moral lightning-conductors' which could shield us from envy, calumny, perfidy and fraud, from egotism and mediocrity, from arrogance and pedantry, and other minor inconveniences. With this object in mind, and to bring the day closer when the entire conduct of the affairs of the world would be mechanically governed and controlled, he proposed the establishment of an Academy of Syllographs, which would offer prizes for the invention of three automata. The first prize would be awarded to the inventor of an automaton to represent a man who would not calumniate his absent friend, defend him when ridiculed in his absence, harbour no envy of him and guard his interests. The same automaton would also have to have the property of never, from mere love of gossip, divulging a secret committed to its care. The second prize would be awarded for an 'artificial steam man' designed to perform acts of virtue and magnanimity, the Academy being of the opinion that the power of steam might prove effective where all other motive forces had failed. In this way an automaton might attain true glory. The third prize would be awarded for an automaton capable of performing the duties of a faithful wife while remaining a source of conjugal bliss, this being a combination of functions seemingly without precedent.

The fundamental question appears to be whether or not we are prepared to recognise a psychological world of meaningful experience, in which a subjective-objective anti-

---

[25] John Cohen, *Behaviour in Uncertainty, op. cit.*, Chap. 6.

[26] J. M. Stroud, 'The Fine Structure of Psychological Time', pp. 174–205 in *Information Theory in Psychology* (ed. H. Quastler), Glencoe, III, The Free Press, 1955. *See also* M. Taube, *Computers and Common Sense*, New York: Columbia University Press, 1961, for a sobering view of the future of computers, especially in relation to translation, and, for a valuable general discussion, F. H. George, *The Brain as Computer*, London: Pergamon Press, 1961.

[27] M. Polanyi, 'The Hypothesis of Cybernetics', *Brit. J. Phil. Sci.*, 1952, 11, pp. 312–15; and for a more extended analysis, *Personal Knowledge, supra* ref. 21.

[28] Vartanian, *op. cit.*, p. 136.

[29] G. Leopardi, *Essays, Dialogues and Thoughts* (transl. by P. Maxwell), London: Scott, 1893.

thesis is false, for the psychological is properly 'objective', and the so-called 'objective' is in the last resort psychological. If we may only, in the name of science, speak of neural events, there is no place for meaning and hence for any concept of 'mind'; with such a solution a robot could be perfectly content.

The Feelings of Robots

by Paul Ziff

Dr. Ziff's argument that robots could not have feelings aroused a lively controversy — as may be seen in the papers by J. J. C. Smart and Ninian Smart that are appended to Dr. Ziff's original article. Here, as in Professor Cohen's commentary and in much of the current and recent writing on the differences between man and automata, the essential distinction appears to be emotion. Mind over matter can be a triumph of the computer as well as man; but only human beings can laugh, weep, love — and blush.

Could a robot have feelings? Some say of course.[1] Some say of course not.[2]

1. I want the right sort of robots. They must be automata and without doubt machines.

I shall assume that they are essentially computing machines, having microelements and whatever micromechanisms may be necessary for the functioning of these engineering wonders. Furthermore, I shall assume that they are powered by microsolar batteries: instead of having lunch they will have light.

And if it is clear that our robots are without doubt machines then in all other respects they may be as much like men as you like. They may be the size of men. When clothed and masked they may be virtually indistinguishable from men in practically all respects in appearance, in movement, in the utterances they utter, and so forth. Thus except for the masks any ordinary man would take them to be ordinary men. Not suspecting they were robots nothing about them would make him suspect.

But unmasked the robots are to be seen in all their metallic lustre. What is in question here is not whether we can blur the line between a man and a machine and so attribute feelings to the machine. The question is whether we can attribute feelings to the machine and so blur the line between a man and a machine.

2. Could robots have feelings? Could they, say, feel tired, or bored?

Ex hypothesi robots are mechanisms, not organisms, not living creatures. There could be a broken-down robot but not a dead one. Only living creatures can literally have feelings.

If I say "She feels tired" one can generally infer that what is in question is (or was or will be in the case of talk about spirits)[3] a living creature. More generally, the linguistic environment ". . . feels tired" is generally open only to expressions that refer to living

From *Analysis*, vol. 19, no. 3 (1959). Reprinted by permission of the editor of *Analysis*.

[1] Cf. D. M. MacKay, "The Epistemological Problem for Automata," in *Automata Studies* (Princeton: Princeton University Press, 1956), pp. 235ff.

[2] Cf. M. Scriven, "The Mechanical Concept of Mind". [In A. R. Anderson, *Minds and Machines* (Englewood Cliffs: Prentice-Hall, 1964), pp. 31ff.]

[3] I shall henceforth omit the qualification.

creatures. Suppose you say "The robot feels tired." The phrase "the robot" refers to a mechanism. Then one can infer that what is in question is not a living creature. But from the utterance of the predicative expression ". . . feels tired" one can infer that what is in question is a living creature. So if you are speaking literally and you say "The robot feels tired" you imply a contradiction. Consequently one cannot literally predicate ". . . feels tired" of "the robot."

Or again: no robot will ever do everything a man can. And it doesn't matter how robots may be constructed or how complex and varied their movements and operations may be. Robots may calculate but they will not literally reason. Perhaps they will take things but they will not literally borrow them. They may kill but not literally murder. They may voice apologies but they will not literally make any. These are actions that only persons can perform: *ex hypothesi* robots are not persons.

3. "A dead robot" is a metaphor but "a dead battery" is a dead metaphor: if there were a robot around it would put its metaphor to death.

What I don't want to imply I need not imply. An implication can be weakened. The sense of a word can be widened or narrowed or shifted. If one wishes to be understood then one mustn't go too far: that is all. Pointing to one among many paintings, I say "Now *that* one is a *painting.*" Do I mean the others are not? Of course not. Yet the stress on "that" is contrastive. So I say "The robot, that mechanism, not of course a living creature but a machine, it feels tired": you cannot infer that what is in question here is a living creature.

If I say of a person "He feels tired," do you think I am saying that he is a living creature and only that? If I say "The robot feels tired" I am not saying that what is in question is a living creature, but that doesn't mean that nothing is being said. If I say "The robot feels tired," the predicate ". . . feels tired" means whatever it usually means

except that one cannot infer that what is in question is a living creature. That is the only difference.

And what has been said about "The robot feels tired" could be said equally well about "The robot is conscious," "The robot borrowed my cat," and so forth.

4. Could robots feel tired? Could a stone feel tired? Could the number 17 feel tired? It is clear that there is no reason to believe that 17 feels tired. But that doesn't prove anything. A man can feel tired and there may be nothing, there need be nothing at all, that shows it. And so with a robot or a stone or the number 17.

Even so, the number 17 could not feel tired. And I say this not because or not simply because there are no reasons to suppose that 17 does feel tired but because there are good reasons not to suppose that 17 feels tired and good reasons not to suppose that 17 ever feels anything at all. Consequently it is necessary to consider whether there are any reasons for supposing that robots feel tired and whether there are good reasons for not supposing that robots ever feel anything at all.

5. Knowing George and seeing the way he looks I say he feels tired. Knowing Josef and seeing the way he looks I don't say he feels tired. Yet if you don't know either of them then to you George and Josef may look alike.

In one sense they may look alike to me too, but not in another. For George but not Josef will look tired. If you ask me to point out the difference there may be nothing relevant, there need be nothing relevant, to point to. For the relevant difference may be like that between looking at an unframed picture and looking at it framed. Only the frame here is provided by what I know about them: you cannot see what I know.

(Speaking with the robots, one can say that the way things look to me, my present output, will not be the same as yours, the way things look to you, even though at present we may both receive the same input, the same stimuli, and this is because your mechanism was not in the same initial state as mine, owing either to a difference in structure or to a difference in previous inputs.)

If we say of a person that he feels tired, we generally do so not only on the basis of what we see then and there but on the basis of what we have seen elsewhere and on the basis of how what we have seen elsewhere ties in with what we see then and there. And this is only to say that in determining whether or not a person feels tired both observational and theoretic considerations are involved and, as everywhere, are inextricably interwoven.

6. Suppose you and I visit an actor at home. He is rehearsing the role of a grief-stricken man. He ignores our presence as a grief-stricken man might. His performance is impeccable. I know but you do not know that he is an actor and that he is rehearsing a role. You ask "Why is he so miserable?" and I reply "He isn't." "Surely," you say, "he is grief-stricken. Look at him! Show me what leads you to say otherwise!" and of course there may be nothing then and there to show.

So Turing[4] posed the question whether automata could think, be conscious, have feelings, etc., in the following naïve way: what test would an automaton fail to pass? MacKay[5] has pointed out that any test for mental or any other attributes to be satisfied by the observable activity of a human being can be passed by automata. And so one is invited to say what would be wrong with a robot's performance.

Nothing need be wrong with either the actor's or a robot's performance. What is wrong is that they are performances.

7. Suppose K is a robot. An ordinary man may see K and not knowing that K is a robot, the ordinary man may say "K feels tired." If I ask him what makes him think so, he may reply "K worked all day digging ditches. Anyway, just look at K: if he doesn't look tired, who does?"

So K looks tired to the ordinary man. That doesn't prove anything. If I know K is a robot, K may not look tired to me. It is not what I see but what I know. Or it is not what I see then and there but what I have seen elsewhere. Where? In a robot psychology laboratory.

8. If I say "The robot feels tired," the predicate ". . . feels tired" means whatever it usually means except that one cannot infer that what is in question is a living creature. That is the only difference.

To speak of something living is to speak of an organism in an environment. The environment is that in which the behavior of the organism takes place. Death is the dissolution of the relation between an organism and its environment. In death I am′ pluralized, converted from one to many. I become my remains. I merge with my environment.

If we think of robots being put together, we can think of them being taken apart. So in our laboratory we have taken robots apart, we have changed and exchanged their parts, we have changed and exchanged their programs, we have started and stopped them, sometimes in one state, sometimes in another, we have taken away their memories, we have made them seem to remember things that were yet to come, and so on.

And what we find in our laboratory is this: no robot could sensibly be said to feel anything. Why not?

9. Because there are not psychological truths about robots but only about the human makers of robots. Because the way a robot acts (in a specified context) depends primarily on how we programed it to act. Because we can program a robot to act in any way we want it to act. Because a robot could be programed to act like a tired man when it lifted a feather and not when it lifted a ton. Because a robot couldn't mean what it said any more than a phonograph

[4] Cf. "Computing Machinery and Intelligence" ([in Anderson, op. cit.,] pp. 4ff.).

[5] Cf. "Mentality in Machines," *Proceedings of the Aristotelian Society*, Supp. Vol. XXVI (1952), 61ff.

record could mean what it said. Because we could make a robot say anything we want it to say. Because coveting thy neighbor's robot wife would be like coveting his car and not like coveting his wife. Because robots are replaceable. Because robots have no individuality. Because one can duplicate all the parts and have two virtually identical machines. Because one can exchange all the parts and still have the same machines. Because one can exchange the programs of two machines having the same structure. Because. . . .

Because no robot would act tired. Because a robot could only act like a robot programed to act like a tired man. For suppose some robots are programed to act like a tired man after lifting a feather while some are so programed that they never act like a tired man. Shall we say "It is a queer thing but some robots feel tired almost at once while others never feel tired"? Or suppose some are programed to act like a tired man after lifting something blue but not something green. Shall we say "Some robots feel tired when they lift blue things but not when they lift green things"? And shall we conclude "Some robots find blue things heavier than green things"? Hard work makes a man feel tired: what will make a robot act like a tired man? Perhaps hard work, or light work, or no work, or anything at all. For it will depend on the whims of the man who makes it (though these whims may be modified by whatever quirks may appear in the robot's electronic nerve network, and there may be unwanted and unforeseen consequences of an ill-conceived program). Shall we say "There's no telling what will make a robot feel tired"? And if a robot acts like a tired man then what? Some robots may be programed to require a rest, others to require more work. Shall we say "This robot feels tired so put it back to work"?

What if all this were someday to be done with and to human beings? What if we were someday to break down the difference between a man and his environment? Then someday we would wake and find that we are robots. But we wouldn't wake to a mechanical paradise or even an automatic hell: for then it might not make sense to talk of human beings having feelings just as it now doesn't make sense to talk of robots having feelings.

A robot would behave like a robot.

Professor Ziff on Robots

by J. J. C. Smart

Professor Ziff ("The Feelings of Robots") argues that robots could not have feelings. Only living things, he says, could have feelings, and robots could not be living things. Both his premise and his conclusion seem to me to be questionable, though in a note I can touch on some only of my reasons for thinking this.

(A) The notion of "living thing" as opposed to "robot" is unclear.

(1) Let us pretend that the *Genesis* story is literally true. Then Adam and Eve were robots. They were artifacts fashioned by God. If a conflation of ancient theology and modern biology may be allowed, we could even say that God gave Adam and Eve "programs," namely their sets of genes, probably DNA molecules which have the function of recording hereditary information.

From *Analysis*, vol. 19, no. 5 (1959). Reprinted by permission of the author and the editor of *Analysis*.

Robot pilot, c. 1925. Professor J. Popjie settles into the cockpit of his airplane while a mechanical man of his own invention prepares to operate the controls in a flight over Brighton, England. (New York Public Library, Picture Collection)

(2) Consider von Neumann's self-reproducing mechanism. (John von Neumann, "The General and Logical Theory of Automata," *Cerebral Mechanisms in Behaviour*, The Hixon Symposium, 1951, pp. 1–31.) In what sense would descendants of such a mechanism be any the less living creatures than descendants of Adam and Eve? We could even suppose small random alterations in that part of them which records their design. Such machines could evolve by natural selection and develop propensities and capacities which did not belong to the original machine.

(1) and (2) taken together show how unclear is the distinction between a (sufficiently complex) artifact and a living creature. I myself find this not in the least surprising, for I am inclined to accept the physicalist thesis that living creatures just are very complicated physico-chemical mechanisms.

(B) I cannot see why "this has feelings" entails "this is a living creature," if "this is a living creature" is taken to entail "this is not an artifact." None of the artifacts any of us have met in practice have been sufficiently complex to warrant the assertion that they have feelings. So we should in fact at present run into no trouble if we deduced "this is a living creature" if told "this has feelings." Nevertheless it need not be a logical entailment. Maybe in the future we shall find counterexamples.

(C) Suppose we made a robot so complex that it could learn new purposes and capacities in the way that a child can. (Compare Turing's child-machine, in "Computing Machinery and Intelligence." It might even become a philosopher, attending conferences, and developing just as human philosophers do. Why should we not say that it *meant* what it said? It would not be at all analogous to Ziff's machine with a phonograph record inside.

In short, therefore, I find Ziff's arguments unconvincing. I suspect that I may have misunderstood their purport, but I feel that Ziff could have made his intentions plainer.

Robots Incorporated

by Ninian Smart

Paul Ziff, discussing robots . . . , gives a number of reasons why it is not to be said that they feel tired and so on. I shall urge that his reasons would not persuade a determinist, except insofar as he is bringing out this point: that his robots do not feel because they do not have proper bodies. I shall then argue that this does not help much to see why certain artifacts do not feel. One does not, of course, need to be a determinist; but if one is not, and unless unrealistically one is going to compartmentalize human beings, the reasons why human beings feel tired and robots do not must go much deeper. I shall not deal with all Ziff's reasons, but will concentrate on certain of them to be found in Section 9 of his article, which I take to be crucial. Robots, he argues do not feel because: —

1. "The way a robot acts (in a specified context) depends primarily on how we programed[1] it to act." For the sake of simplicity I introduce the notion of Nature to represent the sum of causes going towards the creation of a human being considered as beginning with conception or at any later time in his life. What is wrong, for the determinist, in saying that the way a man acts, in a specified context, depends primarily on how Nature programs him to act? Subtle programs, of course; much subtler than computer programs, but the subtle cell circuits still determine the way I act, given a situation.

2. "We can program a robot to act in any way we want it to act." Nature does not want things, so in a way the parallel does not hold. But this is a side issue: for the status of an effect need not be affected by the status of the cause. Is Ziff then bringing out our sense of power over robots? Not quite, for we might gain what appears to be a similar power over the animal kingdom through crafty breeding (give me a cat that loves mice — but don't think you are giving me a cat that cannot feel). No, the special sort of power is brought out as follows:

3. "Because a robot could be programed to act like a tired man when it lifted a feather and not when it lifted a ton." The power is that the same structure can be programed opposite ways, not that different robots can be built that operate oppositely (we cannot breed a cat, perhaps, that loves mice until we change its program). For the determinist, the trouble is that Ziff's robot does not possess a proper body: it is not fully incorporated. Nature gives the cat its program (or set of programs) with its body, and for a new program you need a new body.

4. "You can duplicate all the parts." Nature replaces all or most of the parts every so often. And Nature produces identical twins. For that matter Nature might produce centuplets — it is a dull but logically possible world where human beings roll out of the womb with the regularity and indistinguishability of Cadillacs.

From *Analysis*, vol. 19, no. 5 (1959). Reprinted by permission of the author and the editor of *Analysis*.

[1] In what follows, both Ziff and I use "program" in a rather wide sense.

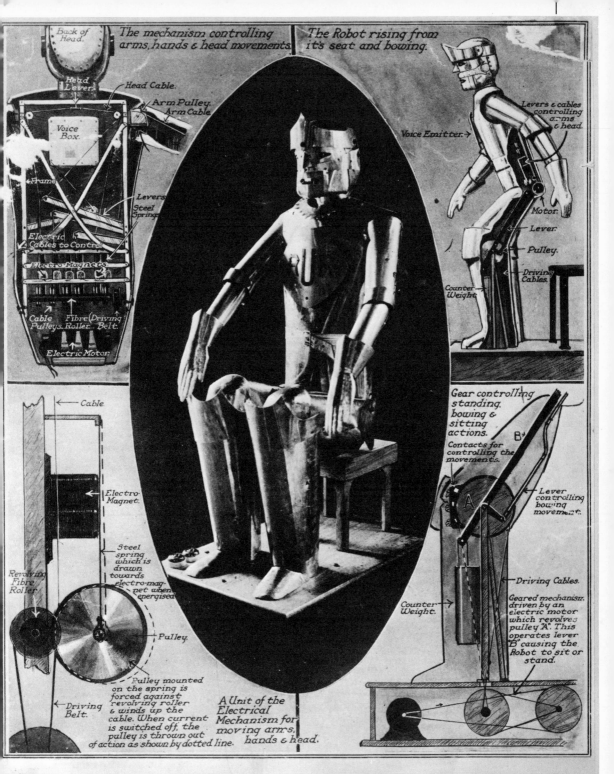

AN ALUMINIUM "MAN" THAT RISES, BOWS, AND MAKES A "SPEECH" : A KNIGHT-LIKE ROBOT.

There has just been completed at Gomshall, near Dorking, the first British Robot, a gleaming thing of aluminium, not unlike a mediæval knight in armour, whose first duty will be to open a Model Engineering Exhibition to-day at the Royal Horticultural Hall. Concealed in the body is an electric motor which drives a fibre roller. Just above are several electro-magnets, with steel springs. To the base of these springs are fixed pulleys carrying cables that operate levers which move the Robot's arms and head. When the electro-magnets are energised the springs are drawn to the magnets pulling the edges of the pulleys against the revolving fibre roller. The pulleys revolve, winding in the cable and moving the head or limbs as desired. By cutting off current the wheel-face is detached from the roller, and the arm falls back to its normal position. For raising the Robot from its seat, causing it to bow to the audience and resume its seat, another motor is concealed in the platform below the figure's feet. This operates large pulley wheels concealed in the knees. When these wheels are slowly turned, a lever attached to each raises or lowers the man as required. Three contacts on the pulley give the desired positions to the operator. A second lever tilts the body and gives the "bowing" movement. To ease the work of the motor, counter-weights in the legs balance the weight of the body and interior mechanism. An ingenious electrical gear (which is the jealously guarded secret of the inventors) enables the Robot to hear questions and answer in a human voice. The Robot has been designed and made in under six months, so that it is but an infant and not yet able to walk, but the inventors state that in time it will be able to use its legs. At present, however, its chief work will be in the realms of publicity.

(*From the* Illustrated London News, *September* 1928. New York Public Library, *Picture Collection*)

These examples seem to show that, for a determinist reading Ziff, it is the body that counts. Ziff says at the outset that he wants the right kind of robot — one which is essentially a computing machine.[2] And such a one he contrasts with a living organism, which is separate from its environment and where death is a merging into the environment.[3] Briefly: his robot is where you have the same valves and different programs; the organism is where you have the same body and the same program (or set of programs).

Embodiment, then, comes essentially to this: a robot is embodied if and only if, in order to reprogram it, it is necessary both to dismantle it and to rebuild it in a different way. This is sufficient to accommodate Ziff's crucial point 3 above. But if Ziff can conceive of a robot which can go through the motions of being tired, you can also conceive of a more inefficient specimen which could not be programed to act tired when lifting a feather. But do we want to say that this one feels tired sometimes?

How then are we to pinpoint the difference between embodied robots which do not feel and those embodied beings which do? The only suggestion that springs readily to mind is that it has something to do with subtlety and complexity. Certain mechanical toys are embodied in the above sense, but they are crude. But the criteria of subtlety and complexity are too vague to be useful. And in any case we have a lurking feeling about ghosts. A Ziff robot does not feel because it is a machine in a ghost, not because there is no ghost in the machine. But to be told that robots do not feel because they do not have bodies, though it makes us pause to cogitate, leaves us unhappy too.

[2] p. 151.
[3] p. 153.

Robots

by Paul Brodeur

A glimpse of what can be seen outside a modern robot factory may persuade us that there really is something in life after all — even if it's only a swan song.

It has been more than forty years since the Czechoslovakian playwright Karel Čapek coined the word "robot" and predicted the overthrow of mankind by automatons in a play entitled "R.U.R." (Rossum's Universal Robots), and though Capek's creepy forecast of a robot take-over seems highly unlikely at the present time, his vision of a burgeoning robot population is about to become a reality, as we discovered last week when we drove up to Danbury and visited Unimation, Inc., makers of an industrial robot called the Unimate. The Unimation factory is a one-story concrete-block building in the countryside on the eastern outskirts of Danbury. Its setting is enhanced by a background of lush green hills, by an expanse of lawn bordered by a split-rail fence, and by a small pond that contains a pair of swans. We arrived at ten in the morning and were greeted at the entrance to the factory by Torsten H. Lindbom, who is general manager of the company and one of the developers of the Unimate industrial robot. Mr. Lindbom is a handsome Swede in his middle forties, with blue Scandinavian eyes, black hair, and a serious manner, and when he led us inside the factory the quiet environment of the country was suddenly replaced by a steady droning sound coming from the rear of

the building. "What you hear is a bunch of Unimates being put through their paces," he explained. "Before we take a look at them, however, I'd like to tell you a bit about industrial robots in general and the development of the Unimate in particular."

Mr. Lindbom went on to say that an industrial robot can be defined as a machine with sufficient intelligence to function without continuous human attention, and that it is designed to replace human operators in hot, hazardous, or tedious jobs, such as forging, die casting, plastic molding and spot welding. "An industrial robot must have a memory system that can be programmed for a specific job, a built-in capability to give and receive commands, and an arm assembly and a hand assembly that have five degrees of freedom," he told us. "The Unimate robot has a seven-and-a-half-foot hydraulically powered arm assembly and a pneumatically powered hand assembly, which are operated by servo drives. The arm can swing to the left or right, move up and down, and go in and out, and the hand can swivel around its own axis and bend at right angles to the axis of the arm. The robot is controlled by an electromagnetic memory drum that can be taught a predetermined program of up to two hundred sequential commands. It was invented by George C. Devol, of Greenwich, who took out patents on it in 1954, and it was developed and perfected

159

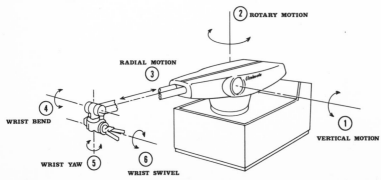

The robot in mass production: Unimate robots spot welding automobile bodies at the Fiat assembly plant in Italy. The diagram illustrates the six program-controlled articulations of these industrial robots. (Courtesy Unimation Inc., Danbury, Conn.)

② ROTARY MOTION

RADIAL MOTION
③

④
WRIST BEND

①
VERTICAL MOTION

WRIST YAW ⑤   ⑥
WRIST SWIVEL

by us over the next ten years. By mid-1962, we had built three Unimate prototypes and put two of them to work in industry. One Unimate unloaded a die-casting machine at an automotive plant in New Jersey, another loaded a press for an electrical manufacturer here in Connecticut, and both performed in a highly successful fashion. From that point on, we were in business. Our production currently stands at four Unimates a month, but when our retooling program is completed we'll have the capacity to produce forty machines a month. And that's only the beginning. We estimate that by 1972 we'll be making a hundred robots a month, and that there will be a total robot population in the United States of at least five thousand."

At this point, Mr. Lindbom led us into a large work area, at the rear of the plant, that contained a dozen or so Unimates, whose housing and arm assembly resemble the turret and cannon of a midget tank. Some of the robots were being assembled by workmen in the center of the room; others were standing off to the side in testing areas, and these, droning away at top speed, as the clawlike hand assemblies were swivelling and bending, looked like quintuple-jointed monsters in convulsion. Mr. Lindbom explained that each of the Unimates was being put through a two-hundred-hour test program prior to being shipped out to its customer; then he stepped up to the control panel of the nearest Unimate, and stilled the robot by turning a switch. Afterward, he picked up an instrument that looked like a push-button telephone, and plugged its cord into a socket on the arm of the Unimate. "This is called the teach-cable assembly and is used to program the robot," he told us. "By pushing buttons on the teach-cable selector panel, we activate the servo drives of the Unimate and lead the robot in slow motion through whatever series of its five movements we wish it to perform. These movements are encoded in binary digital form by five position indica-

tors, called encoders, and are then transmitted to the electromagnetic memory drum. When the desired series of movements has been gone through completely, the drum contains a sequential magnetic memory of the whole process. In the playback cycle, the memory drum returns its instructions to the encoders, which then proceed to serve as feedback devices activating the servo drives that operate the robot's arm and hand. Suppose you want the robot to pick up a hot part from a die-casting machine, carry it through a quenching bath, place it on a cutting mold, and then put it on a conveyor belt. Well, once that program has been recorded in its memory drum, Unimate will repeat the work cycle automatically and endlessly until it is shut off."

Mr. Lindbom went on to tell us that the hand assembly of Unimate provides for interchangeable pairs of fingers, or claws, that can grasp and hold a wide variety of objects, and that a pressure regulator on the hand assembly can be adjusted to enable the robot to pick up an egg as well as a hundred-pound weight. "No question about it, Unimate is a remarkable creature," he said. "It is stronger than a human, it can reach farther, it can withstand hostile environments, and it can remain in one place without getting tired or bored. We've even been teaching it how to putt a golf ball. Two things that Unimate lacks, of course, are eyes and judgment. Since it doesn't have eyesight, it can't locate objects that are not presented to it in orderly fashion, but we're working on a plan to equip it with a television scanner that will feed data into a computer, which, in turn, will orient the robot for sensory contact. As far as judgment goes — that's really something for the future."

At this point, Mr. Lindbom smiled, as if in anticipation of what the future might bring; then he led us through a door at the rear of the factory and into an alley between the main building and a maintenance shop. "The swans in the pond out front made a nest this year, and the female laid five eggs," he told us. "We lost four of the eggs — to dogs and kids, we think — so we brought the fifth out here and hatched it ourselves." He pointed toward the end of the alley, where a fluffy white cygnet sat in a fenced-off area in a patch of shade cast by a homemade coop. "Our maintenance man made the coop," Mr. Lindbom said.

"He takes the cygnet home with him on weekends, so it will be safe. We feed it watercress. When it grows big enough, we'll return it to its parents in the pond." For some moments, Mr. Lindbom gazed down at the tiny cygnet. Then, speaking over the drone of the robots coming through an open window of the factory, he said, "Isn't it beautiful?"

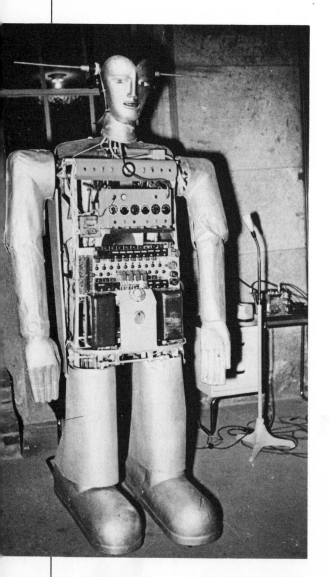

Robot built by Swiss engineer Paul Waltensperger. It is said to be capable of doing "everything except make love" — but before writing that off perhaps one should consult the opinion of a robotess. (The Bettmann Archive, Inc.)

Robots in the Nursery

by Arthur C. Clarke

Arthur C. Clarke speculates on the psychological problems that may be created when the electronic animals of today become the nursery toys of the future.

It is through his toys that a child learns to understand the world in which he lives, for every age is mirrored in miniature on the nursery floor. Many of today's toys — the bulldozers, jet planes, rocket launchers and satellites — would have been unimaginable fifty years ago.

In another half century, our great-grandchildren will be amusing themselves with still more fantastic playthings, the most remarkable of which may be the quasi-intelligent offspring of today's giant computers. For the word "computer" now has a multitude of meanings; it embraces much more than machines which can calculate at lightning-like speed according to some pre-arranged plan. It also describes electronic animals.

The first of these creatures was "born" on the workbench some ten years ago. With the shape, size and speed of a small tortoise, Dr. William Grey Walter's *Machina speculatrix* roamed around the house of one of Britain's leading neurophysiologists, showing complicated and unexpected patterns of behavior despite the fact that its tiny brain contained only two cells, as compared with the ten thousand million inside a human skull. It was no mere robot set to perform some definite task, and it was often impossible to deduce, from its actions alone,

Reprinted from *Holiday Magazine* © 1958 The Curtis Publishing Company.

that *M. speculatrix* was an artificial and not a natural creature. Indeed, it was brighter than many animals, for it could recognize itself in a mirror; and it would bob about in front of the glass until it grew bored and set off again on its endless tour of exploration.

At about the same time, the mathematician Claude Shannon (then at the Bell Telephone Laboratories) built an electronic mouse which could find its way out of a maze by trial and error and then, having performed the feat once, would repeat it indefinitely without going up any blind alleys. This was probably the first machine that could *learn by experience*; it discovered things by making mistakes (don't we all?) but it never made them twice.

Both Grey Walter's and Shannon's primitive pets may seem trivial compared with the giant computers which are taking over so many tasks of the modern world. But they mark the beginning of a new evolution of beings, who some decades hence will share the world with man, either as servant or master.

Soon these electronic animals will have memories and vocabularies superior to those of any living creature. Being able to learn, they will grow up mentally as human children do, storing their experiences and acquiring personalities in the process. Indeed, the best way to teach them would be for man and machine to mature together, though the machine could start with the advantage of having a vast store of information taped into its memory banks.

Such a machine may be the central toy in the nursery of the future. It would be a kind of robot companion, matching the intelligence of the growing child, talking to him so that he learned to use and pronounce words correctly, and teaching him the factual knowledge which parents so often fail to provide.

What would it look like? Certainly not remotely human, for the beauty of the machine seldom mirrors that of the animal world. It might be about the shape and size of a child's pedal car, perhaps with a central turret to carry its various sense organs.

Balloon tires would give it mobility, but it would probably be unable to negotiate a flight of stairs. It would also be quite hope-less at climbing trees; but what boy ever thought badly of his dog because it shared this deficiency?

The nursery robot will combine many of the functions of pet animal, older brother, Mammy (Old Southern style) and school-marm. It could not (or should not) replace the love of parents, but it might often be better than no parents at all. And it will undoubtedly bring into the world a whole host of psychological problems. The attachment of a little girl to her doll, or a boy to his puppy, would be as nothing compared with the affection which a child could develop toward a friendly and intelligent machine.

And so we move into a future of which Freud could never have dreamed; for it is in the electronics labs of today that the complexes of our grandchildren are now being designed.

In Praise of Robots

by Carl Sagan

Astroscientist Carl Sagan argues that man's survival depends to a considerable degree on maximizing the cooperation between human and machine intelligence. Unfortunately, at present, there is a tendency for human beings to underrate the capabilities of robots and computers. Their present limitations are frequently contrasted unfavorably with man's greatest achievements. But since machine intelligence is merely at the beginning of its evolution, such contrasts are equivalent to measuring the mind of an insect against that of Albert Einstein. Sagan endeavors to redress the balance by describing some remarkable achievements and potentialities of existing robots. He predicts that the next generation of intelligent machines will enlarge the separate functions of two different kinds of robots: those with embodied intelligences and those controlled by remotely located intelligences. The latter will increasingly make it possible for man to control and utilize resources and environments that are physically inaccessible to human beings. The development of a great range of "intelligent machines capable of performing tasks too dangerous, too expensive, too onerous, or too boring for human beings" is imminent. Sagan insists that our continued existence as a species depends first on accepting the fact that there is nothing "inhuman" about machines demonstrating superiority to man in certain areas of thought or activity, and then learning to live with that fact in the most humanly constructive ways.

The word "robot," first introduced in the 1920s by the Czech writer Karel Čapek, is derived from the Slavic root for "worker." But it signifies a machine rather than a human worker. Robots, especially robots in space, have lately been getting a bad press. We have read that a human being was necessary to make the terminal landing adjustments on Apollo 11, without which the first manned lunar landing would have ended in disaster; that a mobile robot on the lunar surface could never have been so clever as the astronauts in selecting samples to be returned to earth-bound geologists; and that machines could never have repaired, as men did, the sunshade that was so vital for the continuance of the Skylab missions.

All these comments turn out, naturally enough, to have been written by humans. I wonder if a small self-congratulatory element, a whiff of human chauvinism, has not crept into these judgments. Just as whites can sometimes detect racism and men can occasionally discern sexism, I wonder whether we cannot here glimpse some comparable affliction of the human spirit — a disease that as yet has no name.

Inside the illustration:

PROFESSOR BUTTS TAKES A DRINK OF
STRANGE GIN AND EVOLVES AN INVENTION
FOR OPENING THE GARAGE DOOR WITH-
OUT GETTING OUT OF THE CAR.
DRIVE AUTO BUMPER(A) AGAINST MALLET
PUSHING IT DOWN AND EXPLODING CAP(C)
FRIGHTENING RABBIT(D) WHO RUNS TO-
WARD HIS BURROW(E) PULLING STRING
(F) WHICH DISCHARGES PISTOL(G), THE
BULLET PENETRATES CAN(H) FROM WHICH
THE WATER DRIPS INTO AQUARIUM (I). AS
THE TIDE RISES IN THE AQUARIUM IT
ELEVATES THE FLOATING CORK UP-
RIGHT(J) WHICH PUSHES UP END OF
SEE-SAW (K) CAUSING FLEA(L) TO LOSE
IT'S BALANCE AND FALL ON GEDUNK
HOUND'S TAIL (M) WHO WAKES UP AND
CHASES HIS TAIL ROUND AND ROUND
CAUSING PLATFORM(N) TO SPIN AND
TURN ON FAUCET(O). WATER RUNS
THROUGH HOSE(P) STARTING REVOLVING
LAWN SPRINKLER(Q) ON WHICH ROPE
(R) WINDS ITSELF OPENING GARAGE
DOOR.
OF COURSE, IF YOU WISH, YOU CAN
DRIVE RIGHT THROUGH THE DOOR
AND THEN THERE WON'T BE ANY
OBSTRUCTION LEFT TO BOTHER
YOU IN THE FUTURE.

Professor Butts' Automatic Garage Door Opener, 1928, by Rube Goldberg. (Copyright © Rube Goldberg. Permission granted by King Features Syndicate, Inc., 1928)

The word "humanism" has been preempted by other and more benign activities of mankind. From the analogy with sexism and racism I suppose the name for this malady could be "speciesism" — the prejudice that there are no beings so fine, so capable, and so reliable as human beings.

This is a prejudice because it is, at the very least, a prejudgment — a conclusion drawn before all the facts are in. Such comparisons of men and machines in space are comparisons of smart men with dumb machines. We have not asked what sorts of machines could have been built for the thirty or so billion dollars that the Apollo and Skylab missions together cost.

Each human being is a superbly constructed, astonishingly compact, self-ambulatory computer — capable on occasion of independent decision making and real control of his or her environment. But there are serious limitations to employing human beings in certain environments. For example, without a great deal of protection, human beings would be inconvenienced on the ocean floor, the surface of Venus, the deep interior of Jupiter, or even on long space missions. Perhaps the only interesting information from Skylab that could not have been obtained by machines is that when human beings remain in space for a period of months, they undergo a spectacular loss of bone calcium and phosphorus. This seems to imply that human beings will be incapacitated under zero gravity on missions of six to nine months or more. The minimum interplanetary voyages have characteristic lengths of a year or two. Spinning the spacecraft can produce a kind of artificial gravity, but it is inconvenient and costly.

Because we value human beings highly, we are reluctant to send them on extremely risky missions. If we do send human beings to exotic environments, we must also send along their food, their air, their water, amenities for entertainment and waste recycling, and companions. Machines, however, need no elaborate life-support systems, no entertainment, no companionship, and we do not yet feel strong ethical prohibitions against sending them on one-way or suicide missions.

Certainly for simple missions, machines have proved themselves many times over. Unmanned vehicles performed the first photography of the whole earth and the far side of the moon, the first landings on the moon and Venus, and in the Mariner 9 mission to Mars, the first thorough reconnaissance of another planet. Here on earth it is increasingly common for high-technology manufacturing — in chemical and pharmaceutical plants, for example — to be performed largely or entirely under computer control. In all these activities machines are able, to some extent, to sense errors, correct mistakes, and alert distant human controllers about perceived problems. The powerful arithmetical abilities of computing machines — now hundreds of millions of times faster than unaided human beings — is legendary. But what about really complex discussions? Can machines, in any sense, think through a new problem? Can they make discussions of the branched-contingency-tree variety, which we think of as characteristically human? (That is, I ask question 1; if the answer is A, I ask question 2; if the answer is B, I ask question 3; and so on.)

Some decades ago the English mathematician Alan Mathison Turing described what would be necessary for him to believe in machine intelligence. The condition was simply that he could be in teletype or typewriter communication with a machine and be unable to tell that it was not a human being. Turing imagined the following conversion between a man and a machine:

Q. In the first line of your sonnet, which reads "Shall I compare thee to a summer's day," would not "a spring day" do as well or better?
A. It wouldn't scan.
Q. How about "a winter's day"? That would scan all right.
A. Yes, but nobody wants to be compared to a winter's day.
Q. Would you say Mr. Pickwick reminded you of Christmas?
A. In a way.
Q. Yet Christmas is a winter's day, and I do not think Mr. Pickwick would mind the comparison.
A. I don't think you're serious. By a winter's day one means a typical winter's day, rather than a special one like Christmas.

No machine of this sophistication has yet been built, but so far only a limited amount of effort and money has been put into artificial intelligence, and there are only a half-dozen centers of such activity in the world. One of the more striking results, obtained in a very limited universe of discourse — that of children's blocks — comes from work done at the Artificial Intelligence Laboratory of the Massachusetts Institute of Technology. Here is a dialogue between a man and a machine:

Man. What does the box contain?
Machine. The blue pyramid and the blue box.
Man. How many blocks are not in the box?
Machine. Four of them.
Man. Is at least one of them narrower than the one that I told you to pick up?
Machine. Yes, the red cube.
Man. Is there a steeple?
Machine. Sorry, I don't know the word "steeple."
Man. A steeple is a stack that contains two green cubes and a pyramid, for example.
Machine. I understand.
Man. Are there any steeples now?
Machine. No.
Man. Build one.

The machine does. It reaches out its mechanical hand, moves the cubes, and builds the structure that has just been described.

Man. When did you pick up the pyramid?
Machine. While I was stacking up the red cube, the large red block, and the large green cube.
Man. Why?
Machine. To get rid of it.
Man. Why did you clear off that cube?
Machine. To put it on the large green cube.

As another example, a computer program that imitates a psychiatrist was also once developed at MIT. This is a much simpler program than the blocks program because it does not attempt to "understand" much of what is said to it, but only manipulates the phrases in sentences it has heard in accordance with a contingency tree sensitive to certain words that are considered especially important by humans. This program involves nondirective psychotherapy, which generally speaking, requires extremely simple computer programs. Here is a remarkable dialogue between a human being and a computer psychotherapist. (It is one of the most interesting of such conversations. Many other such machine/human psychiatric dialogues are not nearly so striking. But I suppose most human/human psychiatric dialogues are not very striking either.) In order to keep the respective roles of therapist and patient clear in this interaction, the computer therapist types upper case while the patient types lower case:

men are all alike.
IN WHAT WAY?
they're always bugging us about something or other.
CAN YOU THINK OF A SPECIFIC EXAMPLE?
well, my boyfriend made me come here.
YOUR BOYFRIEND MADE YOU COME HERE?
he noticed that i'm depressed much of the time.
I AM SORRY TO HEAR YOU ARE DEPRESSED.
he's right. i am unhappy.
DO YOU THINK COMING HERE WILL HELP YOU NOT TO BE UNHAPPY?
i need some help, that much is clear.
WHAT WOULD IT MEAN TO YOU IF YOU GOT SOME HELP?
perhaps i could learn to get along with my mother.
TELL ME MORE ABOUT YOUR FAMILY.
my mother takes care of me.
WHO ELSE IN YOUR FAMILY TAKES CARE OF YOU?
my father.
YOUR FATHER?
you are like my father in some ways.
WHAT RESEMBLANCE DO YOU SEE?
you are not very aggressive, but i don't think you want me to notice that.

WHAT MAKES YOU THINK I AM NOT VERY AGGRESSIVE?
you don't argue with me.
WHY DO YOU THINK I DON'T ARGUE WITH YOU?
you are afraid of me.
DOES IT PLEASE YOU TO BELIEVE I AM AFRAID OF YOU?
it pleases my father to think i am afraid of him.
WHAT ELSE COMES TO MIND WHEN YOU THINK OF YOUR FATHER?
bullies.
DOES THAT HAVE ANYTHING TO DO WITH THE FACT THAT YOUR BOYFRIEND MADE YOU COME HERE?

This astonishing — one is very tempted to say perceptive — response from the computer is, of course, preprogrammed. But then, so are the responses of human psychotherapists. No such computer program is adequate for psychiatric use today, but the same can be remarked about some human psychotherapists. In a period when more and more people in our society seem to be in need of psychiatric counseling, and when time sharing of computers is widespread, I can imagine the development of a network of computer psychotherapeutic terminals, something like arrays of large telephone booths, in which, for a few dollars a session, we would be able to talk with an attentive, tested, and largely nondirective psychotherapist. Insuring the confidentiality of the psychiatric dialogue is probably the most important step to be worked out.

Another sign of the intellectual accomplishments of machines is found in games. Even exceptionally simple computers — those that can be wired by a bright ten-year-old — can be programmed to play perfect tic-tac-toe. One computer has played master-class checkers — it has beaten the Connecticut state champion. Chess is a much more difficult game than tic-tac-toe or checkers. Here, programming a machine to win is not easy, and novel strategies have been used, including several successful attempts to have a computer learn from its own experience in playing previous chess games. For example, computers can learn empirically that it is better in the beginning game to control the center of the chess board than the periphery.

So far no computer has become a chess master; the ten best chess players in the world have nothing to fear from any present machine. But several computers have played well enough to be ranked somewhere in the middle range of serious, tournament-playing chess players. I have heard machines demeaned (often with a just audible sigh of relief) because chess is an area in which human beings are still superior. This reminds me of the old joke in which a stranger remarks with wonder on the accomplishments of a checker-playing dog, whose owner replies. "Oh it's not all that remarkable. He loses two games out of three." A machine that plays chess in the middle range of human expertise is a very capable machine; even if there are thousands of better human chess players, there are millions of worse ones. To play chess requires a great deal of strategy and foresight, analytical powers, the ability to cross-correlate large numbers of variables and to learn from experience. These are excellent qualities not only for individuals whose job it is to discover and explore but also for those who watch the baby and walk the dog.

Chess-playing computers, because they have very complex programs, and because, to some extent, they learn from experience, are sometimes unpredictable. Occasionally they perform in a way that their programmers would never have anticipated. Some philosophers have argued for free will in human beings on the basis of our sometimes unpredictable behavior. But the case of the chess-playing computer clearly tells us that, when viewed from the outside, behavior may be unpredictable only because it is the result of a complex although entirely determined set of steps on the inside. Among its many other uses, machine intelligence can help illuminate the ancient philosophical debate on free will and determinism.

With this more or less representative set of examples of the state of development of machine intelligence, I think it is clear that a major effort over the next decade, involving substantial investments of money, could produce much more sophisticated programs. I hope that the inventors of such machines and programs will become generally recognized as the consummate artists they are.

In thinking about the next generation of machine intelligence, it is important to distinguish between self-controlled and remotely controlled robots. A self-controlled robot has its intelligence within itself; a remotely controlled robot has its intelligence located someplace else, and its successful operation depends upon successful communication between its external central computer and itself. There are, of course, intermediate cases in which the machine may be partly self-activated and partly remotely controlled. The mix of remote and *in situ* control seems to offer the highest efficiency for the near future.

We can imagine, for example, such a machine designed for the mining of the ocean floor. Enormous quantities of manganese nodules litter the abyssal depths. They were once thought to have been produced by meteorite infall on the earth, but are now believed to be formed occasionally in vast manganese fountains caused by the internal tectonic activity of the earth. Many other scarce and industrially valuable minerals are likewise to be found on the deep ocean bottoms. We clearly have the capability today to design devices that can systematically swim above or crawl upon the ocean floor, perform spectrometric and other chemical examinations of the surface material, radio back all findings to ship or land,

and by means of low-frequency radio homing instruments, mark the sites of especially valuable deposits.

The radio beacon in the robot will then direct great mining machines to the appropriate locales. The present state of the art in deep-sea submersibles and in spacecraft environmental sensors is compatible with the development of such devices. A similar situation exists for offshore oil drilling, for coal and other mineral mining, and other operations. The likely economic returns from such devices would pay not only for their development but for the entire future space program many times over at the present rate of spending.

Machines can be programmed to recognize particularly difficult situations and to inquire of human operators — working in safe and pleasant environments — what to do next. The examples just given are of devices that are largely self-controlled. Remotely controlled devices are also possible, and a great deal of preliminary work has been done in the remote handling of highly radioactive materials in laboratories of the U.S. Atomic Energy Commission. In the application of this technology, I can imagine a human operator connected by radio link with a mobile machine; for example, the operator is in Manila; the machine is in the Mindanao trench in the Philippine Sea — the deepest level of the earth's surface. The operator is attached to an array of electronic relays that transmit and amplify his movements to the machine and that can, conversely, carry back to his senses what the machine perceives. When the operator turns his head to the left, the television cameras on the machine will turn left, and the operator will see on a great hemispherical television screen the scene revealed by the machine's searchlights and cameras. When the operator in Manila takes a few steps forward, the machine in the abyssal depths will amble a few meters forward; when the operator reaches out his hand, the mechanical arm of the machine will extend

itself. The precision of the man-machine interaction will be such that exact manipulation of material on the ocean bottom by the machine's fingers will be possible. With such devices, human beings will be able to "enter" environments otherwise closed to them forever.

In the exploration of Mars, the time is almost upon us when unmanned vehicles will soft land; only a little further in the future they will roam about the surface of the red planet as some do now on the moon. We are not yet ready for a manned mission to Mars. (Some of us are concerned about such missions because of the danger of carrying terrestrial microbes to Mars and Martian microbes, if they exist, back to earth; and because of the enormous expense involved.)

The unmanned Viking landers scheduled to be deposited on Mars in the summer of 1976, however, have a very interesting array of sensors and scientific instruments. Each lander has two cameras with the resolution and stereoscopic capabilities of the human eye, plus a much greater range of color sensitivity, extending into the near infrared. Each lander has a single ear — a seismometer sensitive only to the lowest frequencies in the audio spectrum. The seismometer is, in fact, equipped with a filter to prevent higher frequencies from being registered. Viking also has an elaborate instrument that is the equivalent of a nose and taste buds. This gas chromatograph/mass spectrometer will enable the landers to detect, in abundances of parts per million, hundreds of different organic molecules, many of which unaided humans are incapable of detecting. The Viking lander sports a single arm with which to collect the Martian soil and withdraw it into its body for further examination. Among the lander's range of instruments are three biology sensors, unknown to human anatomy, that are designed to measure microbial metabolism. (However, were we to ingest soil samples, as Viking will, and were those samples to contain microbes, we might discover them by getting sick.) Although it will have the capacity to taste, Viking will not have to eat while on the Martian surface. The lander is equipped with two radioactive thermal generators, which will generate electricity from the radioactive decay of the element plutonium. This is an energy source that should be adequate for a lifetime of many months on the Martian surface.

(The reprogrammability of computers is an important virtue. Suppose we send a human astronaut to Mars, trained in, say, geology. After he or she lands, we discover that we really should have sent a biologist. Reprogramming the astronaut is equivalent to waiting for that individual to complete, on Mars, a remote postgraduate course in biology. A robot geologist on Mars, however, would require only a few hours to be reprogrammed and debugged for biology.)

The obvious post-Viking approach to Martian exploration, one that takes advantage of the Viking technology, would be a tractor-treaded robot rover that would carry the entire Viking spacecraft slowly over the Martian landscape. But such a vehicle would pose a new problem, one never encountered in machine operation on the earth's surface. Although Mars is the second closest planet to Earth, it is so far away that the light travel time between the two becomes significant. At a typical relative position of Mars and Earth, the red planet is twenty light minutes away. A message from the robot on Mars to the human controller on Earth, traveling at the speed of light, would take twenty minutes to arrive. Thus, if the spacecraft encountered a steep incline, it might send a message of inquiry back to the earth. Forty minutes later the response would arrive, saying something like, "For heaven's sake, stand dead still." By then, of course, an unsophisticated machine would have tumbled into the gully. Any Martian rover consequently requires slope and roughness sensors. Fortunately these are readily available and are even found on some children's toys. When confronted with a steep slope or boulder, a rover so equipped would either stop until it received instructions from the earth in response to its query (sent with a televised picture of the terrain) or back away and move in another and safer direction. If every decision in Martian exploration must be fed through a human controller on the earth, the robot rover can traverse only a few feet an hour. But the lifetimes of such rovers — and of human operators — are so long that a few feet an hour represents a perfectly respectable rate of progress.

Contingency decision networks much more elaborate than anything now in existence can be built into the onboard computers of spacecraft of the 1980s. For objectives more remote than Mars, to be explored further in the future, we can imagine human controllers in orbit nearby. In the exploration of Jupiter, for example, I can envisage operators — installed on a small moon outside the planet's fierce radiation belts — controlling, with only a few seconds' delay, the responses of a spacecraft floating in the dense Jovian clouds or wandering about its metallic hydrogen oceans. Earthbound human beings can also be part of such an interaction loop, provided they have the patience to cope with the increased light travel time involved. As we speculate about expeditions into the farthest reaches of the solar system — and ultimately to the stars — self-controlled machine intelligence will clearly assume heavier burdens of responsibility.

In the development of such machines we find a kind of convergent evolution. The Viking lander is, in a curious sense, like some great, outsize, clumsily constructed insect. It is not yet ambulatory, and it is certainly incapable of self-reproduction. But it has an exoskeleton, a wide range of insectlike sensory organs, and is about as intelligent as a dragonfly. Viking also has an advantage that insects do not; by inquiring of its controllers on earth, it can, in effect, assume the intelligence of a human being —

171

that is, its controllers are able to reprogram the Viking computer on the basis of decisions that they make.

To construct something with the intelligence of an insect may not seem a very impressive feat. But it is a feat that took nature four billion years to accomplish. We have been exploring space for less than a hundred-millionth of that time. A machine of this intelligence is a great human achievement.

As the field of machine intelligence advances and as more and more distant objects in the solar system become accessible to exploration, we will see the development of increasingly sophisticated onboard computers — instruments that slowly climb the phylogenetic tree from insect intelligence to crocodile intelligence to squirrel intelligence and, in the not very remote future, I think, to dog intelligence. Any flight to very great distances must have a computer capable of determining whether it is working properly. There will be no possibility of sending to earth for a repairman. The machine must be able to sense it is sick and skillfully doctor its own illnesses. A computer is needed that is able either to fix or replace its own failed parts or sensors or structural components. Such a computer, which has been called STAR — for Self-Testing And Repairing computer — is on the threshold of development. It employs redundant components, as biology does — we have two lungs and two kidneys partly because each is protection against failure of the other. But a computer can be much more redundant than a human being — we have, after all, only one head and one heart.

In view of the weight premium on deep-space exploratory ventures, there will be strong pressures for continued miniaturization of intelligent machines. Remarkable miniaturization has already occurred; vacuum tubes have been replaced by transistors and wired circuits by printed circuit boards. A few years ago a circuit that occupied much of a 1930s radio set was regularly being printed on the equivalent of the head of a pin. Today the same circuit can be printed on the point of a pin, and the head can accommodate a fair fraction of a small computer.

If intelligent machines for terrestrial mining and space exploratory applications are pursued, the time is not far off when household and other domestic robots will become commercially feasible. Unlike the classical anthropoid robots of science fiction, there is no reason for such machines to look any more human than a vacuum cleaner does. They will be specialized for their functions. There are many common tasks, ranging from bar tending to floor washing, that involve a limited array of intellectual capabilities, although they require substantial stamina and patience. All-purpose ambulatory household robots, capable of performing the domestic functions of a proper nineteenth-century British butler, are probably many decades off. But more specialized machines, adapted to specific household functions, are already on the horizon.

Conceivably, many other civic tasks and essential functions of everyday life could also be carried out by intelligent machines. A recent newspaper report states that gar-

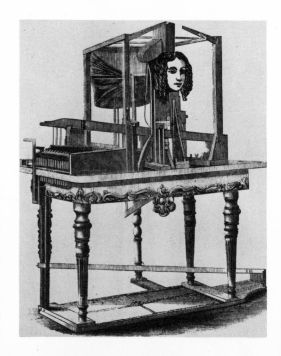

Two talking machines, one in the guise of a Victorian lass (opposite), and the other (above) a friendly automat which said "thank you," possibly making life just a little happier for the lonely late-night diner. (Opposite, Culver Pictures, Inc. Above, Brown Brothers)

bage collectors in Anchorage, Alaska, have won a wage settlement guaranteeing them a yearly salary of $18,000. Economic pressures alone may make a persuasive case for the development of automated garbage-collecting machines. For the development of domestic and civic robots to be a general social good, the effective reemployment of those displaced by robots must, of course, be arranged; but over a human generation that should not be too difficult — particularly if enlightened educational reforms are initiated.

We appear to be on the verge of developing a wide variety of intelligent machines capable of performing tasks too dangerous, too expensive, too onerous, or too boring for human beings. The development of these machines is, in my mind, one of the few legitimate spin-offs of the space program. The main obstacle to their development seems to be a human problem: the quiet feeling that comes stealthily and unbidden to claim that there is something unpleasant or "inhuman" about machines performing certain tasks as well as, or better than, humans; the feeling that generates a sense of loathing for creatures made of silicon and germanium rather than proteins and nucleic acids.

Our survival as a species depends on our transcending these primitive chauvinisms. Adjustment to intelligent machines is, in part, a matter of acclimatization. There are cardiac pacemakers in existence that sense the beat of the human heart. Only at the slightest hint of fibrillation does the pacemaker stimulate the heart. This is a mild but useful sort of machine intelligence. I cannot imagine the wearer of this device resenting its intelligence. I think that there will shortly be a similar sort of acceptance for much more intelligent and sophisticated machines. We have a generation of youngsters who are growing up with pocket computers, machine languages, computer graphics, electronic music, automated instruction, and computer games. They are unlikely to find anything alien about machine intelligence. There is nothing inhuman about an intelligent machine; it is, indeed, the expression of those superb intellectual capabilities that only human beings, of all the creatures on our planet, now possess.

A legitimate concern in the development of machine intelligence is its potential for misuse by unscrupulous governmental, military, and police agencies. Here, as in many other areas of modern technology, the same devices can be used either for enormous good or enormous evil. A world with central data banks containing dossiers on all its citizens, with robot policemen and robot judges, and with automated battlefields is not a world in which I personally would care to live and bring up children. It would be a nightmare world. But a world with adequate food, mineral, and energy resources; a world that provides its human inhabitants with ample leisure and an intellectually and spiritually rich environment with which to make that leisure meaningful; a world engaged in the exploration of other distant and exotic worlds — that is a world I would find extremely attractive. Both of these future worlds are accessible through machine intelligence. To avoid the nightmare and realize the dream requires a wholesale restructuring of the planet's political institutions — a restructuring that is clearly required quite apart from the implications of intelligent machines. If we survive, I think our future will depend to a significant degree on a partnership between human and machine intelligence.

The Vision of John von Neumann

by Adrian Berry

Despite the ubiquitousness of the computer and man's growing dependence on it, there are serious practical and economic limitations to the unrestricted development of artificial intelligence. As Adrian Berry cogently explains, the future of the machine and of man's involvements with it lies not with man-made machines but with automata — i.e., *machine-made* intelligences created by machines that are capable of breeding other machines. The philosophy and methodology of self-reproducing machines was pioneered in the 1940s by Princeton mathematics professor John von Neumann, creator of "games theory" and inventor of MANIAC, the computer that speeded up the making of the first hydrogen bomb. Berry outlines Professor von Neumann's conceptions, explaining, step-by-step, how intelligent machines could reproduce themselves and evolve other, better machines, how these new, artificial intelligences could place man in a position "as sublime as that of the Biblical Jehovah," and how they might ultimately enable man to shape the cosmos to satisfy his needs.

The schemes I have so far described, although modest in scale when measured against the collossal projects considered later in this book, will demand a level of automation far more advanced than today's. To people who know the limitations of modern computers, such instructions to a machine as "Land on the Moon, excavate a cave of such-and-such dimensions, seal its entrances, and construct an airlock" or "Navigate a voyage through superspace to a certain star 33.5 light-years from the sun" will sound unreasonable. Present-day computers, although much more powerful and more compact than the first-generation machines of the fifties, cannot perform such complicated tasks. The most sophisticated machine so far planned is a spaceship computer called ARMMS, initials for the unwieldy title "Automatically Reconfigurable Modular Multiprocessor System." This machine, as its name might suggest after long perusal, is designed to operate and maintain the millions of moving parts of a manned or an unmanned spaceship.[1] It is being specially designed in America for a tour around the giant planets Jupiter and Saturn in the late seventies. It must operate for five years without human assistance. The task of ARMMS would be equivalent on Earth to running a major factory with neither workers nor management, doing all the production, making all the marketing decisions, manipulating

[1] Hughes Aircraft Company Press Release, March 16, 1972.

the prices, and writing the advertisements.[2]

Man-made machines are becoming very much more intelligent, although, as we shall see, the limitation of size will probably make it uneconomic to try to build machines that have the full mental powers of a man. It is now even possible to talk to machines in ordinary English, without enciphering our sentences into ones and zeros. The following conversation by teleprinter between a human and a machine, arranged by scientists at the Massachusetts Institute of Technology, demonstrates the level of intelligence reached by today's most modern machines — and their limitation.[3] The two participants, human and machine, were in different rooms, since the purpose of the test was to learn whether the human *knew* if he was talking to a machine or to a fellow man:

MACHINE. How do you do. I am the doctor. Please sit down at the typewriter and tell me your problem.
HUMAN. I have lost my luggage.
MACHINE. You say you have lost your luggage?
HUMAN. I feel lost without it.
MACHINE. Do you often feel lost without it?
HUMAN. No.
MACHINE. Why not?
HUMAN. Because I normally do not lose things.
MACHINE. What other reasons might there be?
HUMAN. My mother trained me not to lose things.
MACHINE. Tell me more about your family.
HUMAN. I have lost my luggage.
MACHINE. Did your parents ever lose you?

---

[2] Even ARMMS is less sophisticated than STAR ("Self-testing and Repairing Computer"), which would have been built for America's projected "grand tour" of all the solar system's outer planets. STAR would have functioned without human assistance for no less than eleven years. Sadly, all work on it was halted when the grand tour was canceled for budgetary reasons in 1972.

[3] This conversation appears in David Rorvik's *As Man Becomes Machine: The Evolution of the Cyborg* (1971).

Sixty percent of the people subjected to this experiment believed they were talking to a real, if rather stupid psychiatrist in the next room. Indeed, one cannot help feeling that the machine *is* stupid, since it never attempts to consider the all-important problem of how the luggage is to be recovered. Plainly, men are going to build machines very much more capable than this one, even if there is a practical and economic limit, as I will explain later, beyond which artificial intelligence is unlikely to be increased.

That limit is still far ahead. Even household machines are becoming increasingly sophisticated. Kitchen ovens have for several years been available which turn themselves on and off at prearranged times. We can be awakened by a clock device that turns on the radio at any desired hour. Tape recorders built into the telephone system give us information on a widening range of subjects when we dial the appropriate numbers. The computer industry, which was purely experimental in 1945, has now grown so great that according to the predictions of a House of Commons Select Committee in 1971, it will be the world's third-largest single industry in the 1980s, second only to cars and oil.[4] But people are too lazy even to be satisfied with a push-button world. A man presented with this Utopia will ask, "Why can't I have a machine to push the buttons for me?" And so we have invented "command-machines," which, using very small amounts of energy, have the function of controlling much more powerful devices. Well-known examples are the thermostat, the flyball governor on a steam engine, and the photo-electric eye that opens airport doors to approaching passengers. But people, faced with such luxuries, will ask for more: "Why must I go to the trouble of making a thermostat? Why not get a machine to make it for me?" The final step toward real automation will eventually

---

[4] *The Prospects for the United Kingdom Computer Industry in the 1970s.* Fourth Report from the Select Committee on Science and Technology, vol. 1. House of Commons Command Paper 621, November 18, 1971.

be made; it is nothing less than the construction of self-reproducing machines; machines that *breed* machines.

The idea of a supersophisticated machine sounds frightening. Long remembered will be HAL, the talking computer in Stanley Kubrick's film of Arthur C. Clarke's *2001: A Space Odyssey*, which suffered a nervous breakdown because of contradictory orders from its careless human programmers. Instead of protesting to its employer that it was being asked to perform two irreconcilable actions, as a human would have done, it obediently carried out both actions — and then set about murdering all the humans in sight in order to conceal its irrational behavior from the distant programmers. If one machine can avenge a careless error with such perverted and ferocious logic, it is not difficult to imagine the whole world covered with proliferating, murderous machines, engaged through misunderstood instructions on the task of making man extinct. A human seldom needs to be told not to eat the daisies, but a machine, by contrast, sees nothing inherently illogical in committing a lunatic action. Man will therefore find it necessary to protect himself by equipping his machines with safeguard mechanisms. If HAL had been ideally designed, he would have had built-in orders to shut down any group of systems as soon as they began to interact dangerously. He would, at the same time, flash a signal to his programmers saying, "Repair me! I am malfunctioning." Such precautions as these will be essential in the early stages to the scheme for self-reproducing machines worked out in the forties by the celebrated mathematician John von Neumann.

Von Neumann is already well known for other extraordinary contributions to the achievements of our age. Indeed, he possessed one of the most original and penetrating minds of any age. As a colleague remarked:

He functioned magnificently. He had the invaluable faculty of being able to take the most difficult problem, separate it into its components, whereupon everything looked brilliantly simple, and all of us wondered why we had not been able to see through to the answer as clearly as it was possible for him to do.[5]

At the Institute for Advanced Study at Princeton where von Neumann was professor of mathematics from 1933 until his death in 1957, he built MANIAC (short for mathematical analyser, numerical integrator and computer), the most advanced machine of its day, which enabled the hydrogen bomb to be built and tested much sooner than would otherwise have been possible.[6] He invented "games theory," a combination of psychology and mathematical logic, from which have evolved "war games" and those techniques of diplomacy known as "escalation." He suggested also that it was possible for governments to treat economic planning as an exact mathematical science instead of entangling themselves in mazes of ideological doctrine. In a work that may in time be seen to rival the contributions of Keynes, he incorporated entire systems of marketing and production into mathematical equations.[7] He proposed the use of calculus to predict statistically the probability of unforeseen or "irrational" events, such as strikes and wars. In short, he laid the basis for a system of predicting the future over huge time scales. He foresaw that computer technology must grow at a gathering speed. He saw also, as did few others, that man-made computers would be inadequate for extremely complicated tasks. With the experience gained from building MANIAC and several other machines, he was able to astonish an audi-

[5] S. Ulam, "John von Neumann," *Bulletin of the American Mathematical Society*, vol. 64, no. 654, May, 1958. The entire May, 1958, issue of this journal is devoted to discussions by many authors of von Neumann's life's work, edited by B. J. Pettis and G. B. Price.

[6] Because of its grim work, von Neumann's colleagues protested at his plan to name his computer MANIAC. He accordingly referred to it in official literature as JOHNIAC, after himself.

[7] John von Neumann and Oskar Morgenstern, *Theory of Games and Economic Behavior* (1953).

"Arok," a robot invented by Ben Skora of Palos Hill, Ill., looks less like a garbage can than did Dr. Satan's Gort (page 193). One of his useful functions is, in fact, emptying garbage cans. He is also programmed to act as butler, waiter, dog walker, and carpet vacuumer.

"Arok" weighs 275 pounds and is 6 feet 8 inches tall. His master built him from auto parts and household appliances. A modern equivalent of the golem-servant, he would obviously be an asset to any household with a garbage problem. (Ben Skora)

ence in 1948 with a lecture entitled "The General and Logical Theory of Automata."[8]

An automaton, in von Neumann's sense of the word, is a machine wholly constructed by another machine. Its potential abilities are far greater than those of a man-made computer. A man will always find it immensely difficult to build a machine that rivals him in his most innovative qualities. The reason is very simple. His own brain contains about 10 billion nerve and brain cells, or neurons. But these neurons do not *do* things. They only receive information and give orders to the roughly 10,000 *trillion* other cells in the human body. And so we are talking about making mechanical cells numbering about 10 raised to the seventeenth power (i.e., followed by 17 zeros). Manufacturers of computers just do not have the experience to deal with units in such large numbers. Allowing for the utmost miniaturization that may be achieved, a mechanical cell, or integrated circuit (the successor to the transistor), is probably always going to be thousands of times bigger than neurons and body cells. Indeed, because of this difference in size, it can easily be calculated that a machine with the full mental capabilities of a man would need to be almost as large as Westminster Abbey.

Nor is sheer size the only obstacle to the construction of supersophisticated man-made machines. We have seen how HAL's actions became "perverted" because of an undetected programming error. Living creatures are constructed so that malfunctions are as harmless and as inconspicuous as possible. Man-made machines, on the other hand, must be designed deliberately so that their malfunctions are as disastrous and as spectacular as possible. The reason is obvious. Living creatures can operate perfectly well despite minor malfunctions. A scientist with a twisted ankle can still work out a competent theory. But a man-made machine suffering from the electronic equivalent of a twisted ankle cannot possibly be relied on to work efficiently. Any error whatsoever represents a great risk of what von Neumann calls a "generally degenerating process." We would have to inquire urgently: What has caused the malfunction? What other malfunctions is it, in turn, causing? If these questions are not quickly answered, there is a danger that the machine's behavior may go from bad to worse to catastrophic. Repair becomes a far more complicated task than is surgery to living creatures. As von Neumann said in his lecture, "We have to be far more scared by the occurrence of an isolated error and by the malfunction which must be behind it. Our behavior must clearly be that of overcaution, generated by ignorance. Because the possibility exists that the machine may contain several faults, error-diagnosing becomes an increasingly hopeless proposition."[9] And so "Westminster Abbey machines," with integrated circuits representing the equivalent number of neurons and body cells found in a human being, are a technological dead-end. Constructing one would be an act of financial lunacy. Every error discovered would necessitate radical dismantling, and the maintenance costs would therefore be incalculable.

But these difficulties apply only in the case of man-made machines. As we shall see, they need not arise with *machine-made* machines, which are likely to be far more efficient. It must be accepted as self-evident that any system, whether mechanical or organic, is best equipped to reproduce its own kind. It is as idle to expect a group of men to build a perfect machine as to hope that a monkey might give birth to a horse. Major evolutionary changes cannot be expected to occur in a single generation. Men and machines are simply different species at different stages of their evolution.

John von Neumann proposed the con-

---

[8] This famous lecture is reproduced in pages 288–328 of vol. 5 of von Neumann's *Collected Works* (1963).

[9] "The General and Logical Theory of Automata."

struction of machines that would have the same reproductive ability as living organisms; his 1948 lecture showed a new insight into the way in which living creatures reproduce. He lived long enough to see his ideas brilliantly confirmed by the biologists Francis Crick and James D. Watson,[*] who discovered in 1953 that the living cell is made of a very special substance called deoxyribonucleic acid (abbreviated DNA), which carries all the genetic information necessary for the cell's duplication and directs the building of proteins.[10] The DNA does not itself build proteins. It passes instructions for protein-building to another substance with which it is mixed — ribonucleic acid, known as RNA. RNA has been described as the "junior assistant" of DNA. While RNA gets on with the routine and somewhat dull task of building proteins, using for its raw materials those cellular structures known as *ribosomes*, DNA does the really brilliant and imaginative work of programming its *genes*, which decide, in the case of a human baby, whether it shall have dark or fair hair, whether it will grow up short or tall, or whether its temperament will be phlegmatic or excitable. The gene's actual instructions are written on a complicated compound of DNA called *polymerase*.

A computer designer, or even a businessman running an office, may find something familiar about all this, even if he knows nothing of cell structure and has never heard of DNA. For DNA and RNA represent in their respective functions the classic system — any classic system — of efficient organization and expansion. The genius of John von Neumann gave him insight into DNA and RNA long before Crick and Watson proved that they existed. His machines, when they are built and begin to evolve, will contain reproductive systems identical in essence to those of living cells. Like a factory, they must comprise three separate components. In their simplest form, they would be something like this:

Department A, which collects raw materials and processes them in obedience to written instructions.

Department B, a message system, which writes and circulates these written instructions.

Department C, the original author of the instructions, which combines the roles of managing director (who decides what products shall be manufactured), the marketing and sales divisions (who decide what shall be done with the manufactured products), and the company architect (who must design the factory before any of these operations can begin).

It is hardly necessary to repeat that this is just as much a lesson in biology as it is in industrial organization. Polymerase, for instance, fills the roles either of the magnetic tape on a computer or of the notepaper on which the boss writes his memos. In short, Crick and Watson found that the three essential components of the human or animal genetic system were identical to those of von Neumann's automata. Department C is DNA and its assistant RNA, Department A is the ribosomes, and Department B is polymerase.

It seems almost foolish now to ask whether a *machine species* consisting of such components could ever evolve into a civilization. The answer, at the risk of triteness, is that it long ago did so, and that we are that civilization. A living organism, in short, is identical in its basic functions to the automata in von Neumann's universal theory. The fundamental design of every microorganism larger than a virus is, as far as we know, exactly as Von Neumann said it

---

[*] Actually it was Miescher in the 19th century who discovered the presence of DNA in cells. Avery, MacLeod and MacCarthy showed that DNA was the genetic material. The contributions of Crick and Watson were to describe the structure and show how it could both reproduce and convey instructions.— EDS.

[10] Of the millions of words published about DNA and RNA, two of the clearest and simplest accounts appear in John Pfeiffer's *The Cell* (1972) and in Lawrence Lessing's *DNA: At the Core of Life Itself* (1967).

should be.[11] How can a machine reproduce itself? The one way it *cannot* is by receiving the order from its human programmer, "Reproduce yourself." The machine can only reply in effect, "I cannot reproduce myself, since I do not know who or what I am." This approach to the machine would be as absurd as if a man were to give his wife a collection of bottles and flasks containing all the chemical ingredients of one human body and ask her to build a baby. Even if the wife were a brilliant biochemist, she would be unlikely to produce any creature more serviceable than did Baron Frankenstein. A mother reproduces with genes, not with her hands in a laboratory. Instead, the controlling human programmer will perform three simple actions when he sets out to create a dynasty of machines:

1. He gives the machine a complete description of itself.

2. He then gives the machine a second description of itself, *but this second description is of a machine that has already received the first description.*

3. Finally, he orders the machine to create another machine that corresponds precisely to the machine of the second description, and he orders the machine to copy and pass on this final order to the second machine.

Ignoring for a moment the problem of raw materials and power supply, we can see that the man will have created by these actions a self-perpetuating cycle of machines. He has made his Adam and his Eve. Keeping in reserve his essential, self-defensive fourth order ("Destroy all machines answering to the description you now have), he can sit back and allow the machines to carry out whatever tasks he has allotted them. The man's position is now as sublime as that of the Biblical Jehovah. If the second

or third generations of his machines start to "sin," he can destroy them with a "flood" and then start all over again with appropriate improvements. It will no doubt be necessary to perform many wrecking operations of faulty generations of machines, the equivalent perhaps of blasting Sodom and Gomorrah, before really efficient gadgetry can be built. The authors of the Old Testament seem to have had an amazing insight into this aspect of future technology.

Isaac Asimov wrote several ingenious stories about humanoid robots, machines with the exact appearance of human beings.[12] For human safety, their manufacturers were compelled by law to construct each robot so that it obeyed the "Three Laws of Robotics." These were:

1. A robot may not injure a human being, or through inaction allow a human being to come to harm.

2. A robot must obey the orders given it by human beings, except when such orders would conflict with the First Law.

3. A robot must protect its own existence as long as such protection does not conflict with the First and Second Laws.

This ingenious system of machine programming also reminds us of the Old Testament. Asimov's three laws are extraordinarily reminiscent of the fundamental texts in many of the world's religions. "Thou shalt have no other gods before me" is a good example. HAL had evidently not been programmed to refrain from injuring his human "gods." Although there will be no need to build von Neumann machines in the likeness of human beings, it will still be necessary to prevent them from going berserk or putting us out of a job. It will

---

[11] It is debatable whether a virus is a life-form at all. It does not reproduce in either von Neumann's or Crick and Watson's sense of the word, since it borrows the ribosomes from the cells that it invades.

[12] Asimov's novels and collections of short stories about robots are entitled, *I, Robot* (New York: Fawcett World, 1970); *The Rest of the Robots* (New York: Doubleday, 1964); *The Caves of Steel* (New York: Fawcett World, 1972); and *The Naked Sun* (New York: Fawcett World, 1972).

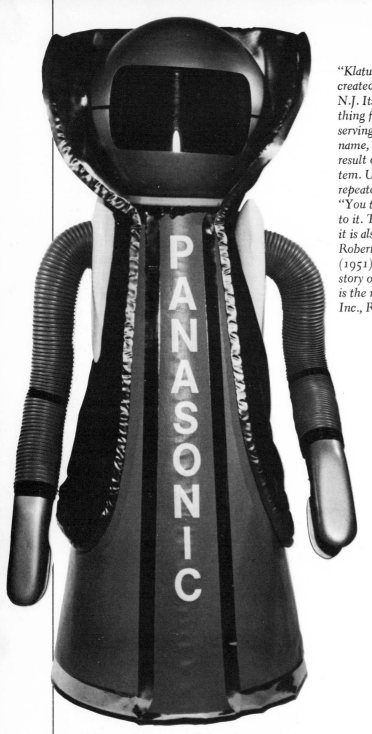

"Klatu" is a 180-pound, 5-foot-2-inch robot created by Anthony Reichelt of Hackensack, N.J. Its inventor claims that it can do "everything from cleaning house and dog-walking to serving drinks and making small talk." Its name, we are told, was bestowed on it as a result of an error in its voice-recognition system. Until the error was rectified, the robot repeated "Klatu" — the phonetic reversal of "You talk" which were the first words addressed to it. This explanation sounds convincing, but it is also noteworthy that the robot-master of Robert Wise's The Day the Earth Stood Still (1951) happens to be called "Klaatu." In the story on which Wise's film was based, "Klaatu" is the robot not the master. (Quasar Industries, Inc., Rutherford, N.J.)

probably be essential to restrict their intelligence to the purely instinctive level.

Professor Freeman J. Dyson, who teaches today at the Princeton Institute for Advanced Study, is a former colleague of von Neumann, and he has given great impetus to the resurrection of his ideas. Dyson is a superb theorist who has worked out engineering schemes on a truly cosmic scale. He believes that self-reproducing machines will be required if man is to generate the colossal amounts of energy necessary for cosmic projects. He gave a rough illustration of how they might be used to solve an economic problem — the comparatively mundane question of how the people of

Earth in the late twenty-first century could obtain their electric power in the improbable event of hydrogen fusion being found to be unpractical.

Dyson suggested to an audience in 1970[13] that the demand for electricity may have become so great by 2070 that nuclear fission power stations line the Pacific coast of America, taking advantage of the heat-absorbing ocean, at an average density of two per mile. These power stations poison the air with their waste discharges, heat up the ocean so that no fish live within a hundred miles, and are of monotonous and hideous appearance. An alternative source of electric power is at last suggested. It is proposed that little colonies of von Neumann machines should be planted in the country's central deserts. After fierce debate, the anti-pollution lobby, which detests power stations, is victorious over the trade union lobby, which supports the continued employment of power workers. The machines are installed. They are fueled by sunlight, and they take moisture from the desert air for their internal needs. They break up rocks to obtain aluminum, silicon, and other minerals with which to construct replicas of themselves. The output of these rock-eating automata is electricity and electrical transmission lines. Proliferating rapidly, they are able in a short time to produce one hundred times more electricity than America consumed in 1970. Unlike the power stations, the machines discharge no waste heat at all, and they create neither smog nor radioactivity.

Dyson himself admits that the obstacles to such a plan are formidable. The machines would certainly require rubber and cadmium to build their transmission lines, which in turn would need some means of traction to carry them from the deserts to the cities. He is not particularly serious in proposing this

idea (believing that the development of fusion will make it unnecessary), and he did so mainly to illustrate his belief that von Neumann machines are best fitted to perform those extremely complicated feats of engineering of which men, using only man-made computers, would be virtually incapable.

I shall pursue no further the future problems of terrestrial power supply, since this book is concerned with the exploitation of space. In the same lecture, Dyson made another, much bolder "thought experiment" involving von Neumann machines. The planet Mars, he pointed out, has at present little economic value since it appears to lack both warmth and water. A group of engineers decide to remedy this without ever leaving their control panels on Earth. A rocket carrying a small colony of von Neumann machines sets out from Earth. It does not head for Mars, but goes instead to Enceladus, a moon of Saturn that consists of mixed rock and water ice. It carries an elaborate program of instructions and some microscopic plant seeds, which will grow in the feeble sunlight falling on Enceladus. The machines will build a greenhouse from local materials for its plants. The plants will, in turn, supply construction materials and fuel to the machines.

To quote Dyson:

For some years after the landing of the rocket nothing unusual seems to be happening to Enceladus. Then, as seen from Earth, Saturn appears to grow a new ring about twice as large as the old rings. A cloud of small objects, each launched from the surface of Enceladus by a simple machine resembling a catapult, begins to spiral slowly outwards from Enceladus's orbit. Each object has a wide, thin sail with which it can navigate in space using the pressure of sunlight.

After another period of years, the outer edge of the new ring extends far out to a place where the gravitational effects of Saturn and the Sun are roughly equal. The small objects come slowly to a halt there, and begin to drift again inwards towards Saturn. This time they are not

[13] Freeman J. Dyson, "The Twenty-First Century." Vanuxem Lecture delivered at Princeton University, February 26, 1970. Unpublished at the time of writing.

moving in spirals any more but in hyperbolic orbits. They zoom past the surface of Saturn at high velocity, receive some last-minute course corrections from computers on Enceladus, and fall free toward the Sun.

A few years later, the night-time sky of Mars begins to glow bright with an incessant sparkle of small meteors. The infall continues day and night, only more visibly at night. Day and night the sky is warm. Soft warm breezes blow over the land, and slowly warmth penetrates into the frozen ground. A few years later, it rains on Mars for the first time. It does not take long for oceans to begin to grow. . . . There is enough ice on Enceladus to keep the Martian climate warm for 10,000 years and to make the Martian deserts bloom.[14]

---

[14] Dyson, "The Twenty-First Century."

This second proposal involving von Neumann machines is very elaborate, and because it requires perfect and very complex behavior by highly advanced machines over several years, during which a single serious malfunction could wreck the whole operation, it is probably fifty years further into the future than Carl Sagan's Venus plan. But we can see perhaps from these ideas that the use of von Neumann machines in some form, the remote descendants of MANIAC, may enable us to alter the environment of the solar system, and even to exploit large parts of the galaxy. Man's activities in the universe must now be envisaged on the truly grand scale.

MOVIES AND MACHINE MEN

"... we only hear machines;
In erg and atom they exact their pay.
And life is largely lived on silver screens ..."
— David McCord

Of all media and art forms cinema is the most obsessively and imaginatively pre-occupied with man's relationships with the machine. Even more than literature it has shaped modern man's fears and fantasies about his own technology. Even more than painting, through a legion of films from Méliès's Clown and the Automaton *(1897) to Kubrick's* 2001: A Space Odyssey *(1968) and the comedic robots of* Star Wars *(1977), it has created the most memorable and sometimes the most disturbing images of our alter egos: the golem, the android, and the computer.*

LEFT: *Charles Ogle as Frankenstein's daemon in the first movie adaptation of Mary Shelley's novel. An Edison film directed by Searle Dawley in 1910, there are no known extant prints of the movie. This still is all that survives to suggest what it was like. (The Museum of Modern Art/Film Stills Archive)*

BELOW: *The best-known movie treatment of synthetic man is Boris Karloff's portrayal of the Monster in James Whale's 1931 film adaptation of Frankenstein. Jack Pierce's admirable makeup job and Karloff's injection of pathos into an almost inarticulate role combined to elevate this interpretation into the standard by which all subsequent Frankenstein Monsters must be measured. (The Museum of Modern Art/Film Stills Archive)*

OPPOSITE: *The Golem (Germany, 1920), directed by Paul Wegener and Carl Boese. Outside the walls of the Prague Ghetto, a little girl (Greta Schröder) fearlessly confronts the rampaging Golem (Paul Wegener) and plucks the Star of David from his chest when he stoops down toward her. The monster is promptly immobilized. (The Museum of Modern Art/Film Stills Archive)*

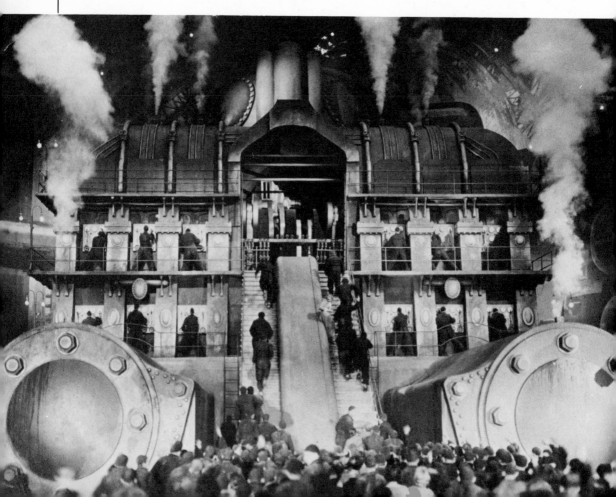

LEFT, TOP: Metropolis. *Tempera poster
by E. McKnight-Kauffer for the British premiere of the film in 1926. (Collection, The
Museum of Modern Art, New York)*

LEFT, BOTTOM: Metropolis. *In the Machine
Shop—the vision of the Pater Noster machine
transformed into Moloch, devourer of the
workers. (Courtesy of National Film Archive/
Stills Library, London)*

RIGHT: Metropolis *(Germany, 1926),
directed by Fritz Lang. Original poster for the
film, depicting Brigitte Helm as the android.
(Private Collection)*

BELOW: Metropolis. *The inventor Rotwang
introduces his robot . . . "Isn't it worth the
loss of a hand to have created the workers of
the future — the machine men?"
(Academy of Motion Picture Arts and
Sciences)*

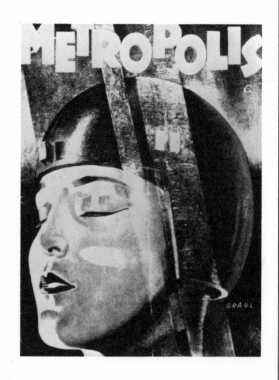

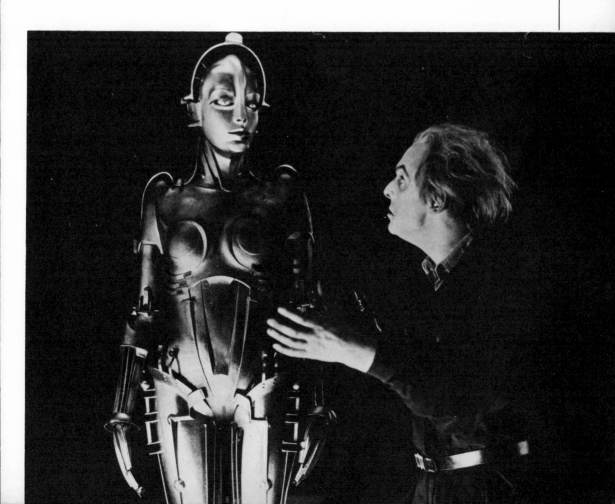

Magnus Zeller, Hitlerstadt *(oil, 1938/39).
Moloch of the Nazi death machine — a striking analogue to the vision of the machine-
turned-Moloch in* Metropolis. *(Märkisches
Museum, Berlin)*

Modern Times (1936), *directed by Charles Chaplin. The heartless machinery of mass production swallows up the Little Fellow — but cannot destroy him. (The Museum of Modern Art/Film Stills Archive)*

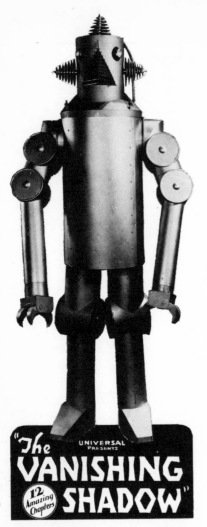

LEFT: The Vanishing Shadow (1934), *a twelve-episode serial directed by Louis Friedlander and released by Universal, featured Onslow Stevens as an intrepid hero avenging his father's death at the hands of a ruthless political faction. He was aided by the marvelous inventions of Professor Carl Van Dorn (James Durkin) which included a death ray and a robot. The latter's appearance was typical of many mechanical men depicted in the pulp fiction and science fiction films of the period. (Copyright © Universal Pictures)*

BELOW: The Birth of the Robot (1936), *directed by Len Lye. A short animated film on the role of automata and mechanization in modern life — by Britain's master of animation in the 1930s. (Courtesy National Film Archive/Stills Library, London)*

RIGHT: *Gort, the robot in* The Mysterious Dr. Satan (1940), *a fifteen-episode Republic serial directed by William Witney and John English. Ultimately, this man of true metal destroys the mysterious doctor (Eduardo Cianelli), thereby frustrating his evil plans for terrorizing America. The robot's name anticipates that of the robot in Robert Wise's classic science fiction film,* The Day the Earth Stood Still (1951), *but there the resemblance ends. Dr. Satan's robot looks distinctly as if he were related to the garbage can in which he finally winds up.* (Authors)

BELOW: The Day the Earth Stood Still (1951), *directed by Robert Wise. Leaving the flying saucer are the robot Gort (Lock Martin), Helen Benson (Patricia Neal), and Klaatu (Michael Rennie). Gort has the power to destroy the earth — unless held in check by the mysterious words: "Klaatu barada nikto!"* (Copyright © 1951 Twentieth Century-Fox Film Corp. All rights reserved. Courtesy of Twentieth Century-Fox.)

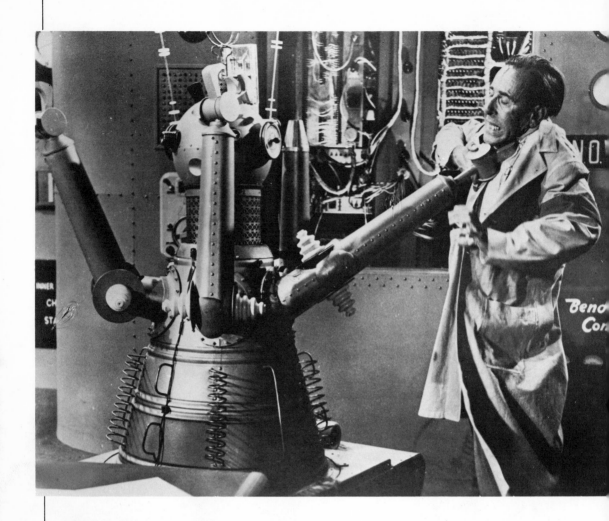

ABOVE: Gog (1954), *directed by Herbert L. Strick. The still shows what happens when the robot Gog goes berserk, after shaking off the restraining power of a benevolently disposed "mechanical brain" called Novac (Nuclear Operative Variable Automatic Computer). Like most of the movie robots of the fifties, Gog is obviously quite unfamiliar with Isaac Asimov's laws of robotics. (Movie Star News)*

LEFT: *One of Robbie the Robot's numerous reappearances following his memorable debut in Forbidden Planet (1956), directed by Fred McLeod Wilcox. (Copyright © 1956 by Loew's Incorporated)*

LEFT: *Dr. Smith (Jonathan Harris) indulges in a not too exceptional altercation with the friendly robot of television's popular series* Lost in Space, *which was based on* Forbidden Planet.

BELOW: 2001: A Space Odyssey *(1968), directed by Stanley Kubrick: Dave Bowman (Keir Dullea) surveys the control panel of the spaceship* Discovery, *in the center of which is the red eye of the malevolent computer, HAL 9000.*

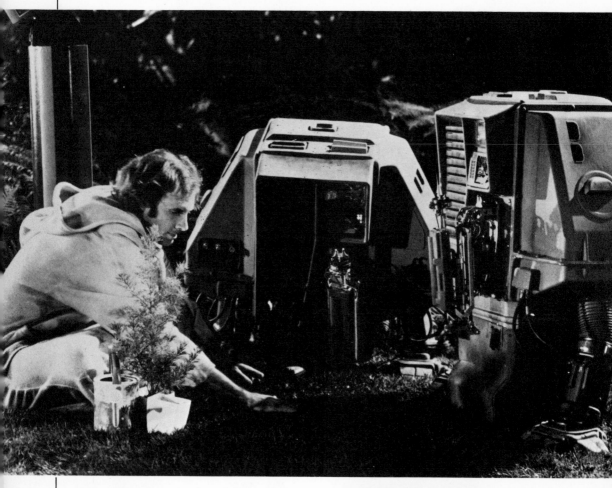

ABOVE: Silent Running (1972), *directed by Douglas Trumbull. Alone with only drones aboard a gigantic space freighter, botanist Lowell Freeman (Bruce Dern) teaches the robots how to plant trees. (Copyright ©️ Universal Pictures)*

RIGHT: Sleeper (1973), *directed by Woody Allen. Having woken up in a hostile world of the future, Woody tries to evade his pursuers by passing himself off as a robot. United Artists. (Authors)*

OPPOSITE ABOVE: Westworld (1973), *directed by Michael Crichton. You can have fun among the androids of Westworld — up to a point. Here the champion gunslinger (Yul Brynner) is being reassembled prior to his deadly pursuit of Richard Benjamin. . . . If the best-laid schemes of mice and men cannot be relied upon, what can one expect of a computer that is supposed to keep its robots under control? (Copyright ©️ 1973 by Metro-Goldwyn-Mayer Inc.)*

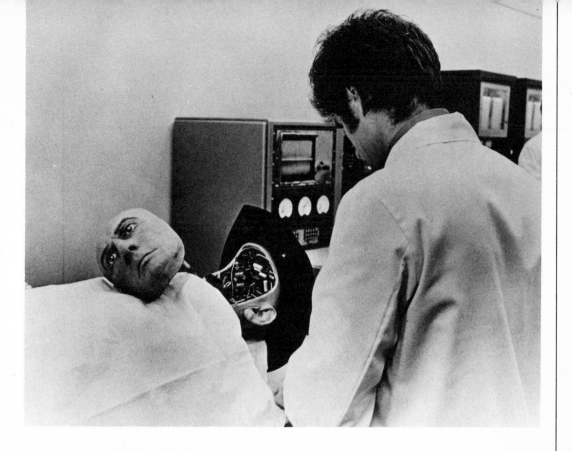

BOTTOM: *Susan Harris (Julie Christie) is paralyzed with terror as a humanoid mechanical arm forces her to do a power-* *hungry computer's bidding in Demon Seed (1977), directed by Donald Hammell. (Copyright © 1977 by Metro-Goldwyn-Mayer Inc.)*

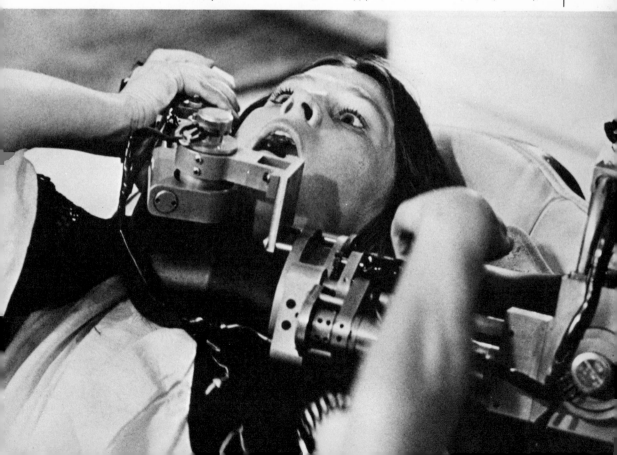

FUTURE VISIONS— FUTURE PROBLEMS:

SCIENCE FICTION AND ROBOTS

There can be no final word on the machine until man himself has dissolved into what Bernard Shaw called "a vortex in pure thought." Ultimate visions are another matter. Here as an envoi are just a few — a small detachment from the endless legions of automata, androids, and robots that march — often menacingly — through the pages of the science fiction pulps and across the covers and jackets of sci-fi novels.

The Coming of the Humanoids

by Neil P. Hurley, S.J.

Father Hurley considers science fiction as a means of confronting ourselves with the implications, probabilities, and possibilities of our technology. He is particularly intrigued with the question "At what point does the human die and factory parts take over?" and his discussion is enlivened with thought-provoking examples from recent fiction about robots, androids, and humanoids.

Science fiction has proven in the last 50 years to have a greater accuracy of prediction than, as a rule, demographers, economists and political forecasters. The reason lies partly in the fact that the authors of science fiction have intuitively grasped the inexorable dynamic behind the mechanization process.

In science fiction we see portrayed the uttermost limits of mechanization's logic, an extreme rationalistic view which arbitrarily arranges parts and techniques into utilitarian patterns without concern for ultimate purposes. Since the all-absorbing process of "automatization" is self-justifying, its inevitable term is a machine which is not a machine. Since Karel Čapek's *R.U.R.* in 1921, dramatists, novelists and short story writers have played fancifully, sometimes frighteningly, with the idea of a robot which could simulate human behavior. In its most perfect incarnation, the robot takes on such life-like appearance as to be called an android or humanoid. The growing spate of fiction on androids is beginning to constitute a new literary category within the larger *genre* of science fiction.

From *Commonweal*, vol. 91, no. 10 (December 5, 1969), pp. 297–300. Reprinted by permission of Commonweal Publishing Co., Inc.

We are all familiar with the heart transplant operation. In his short story, *The Fires of Night*, Dennis Etchison asks about replacing vital organs with, say, a plastic heart, a latex stomach, a fiberglass ribcage. He then poses the next logical question: Is it murder to kill someone who was part flesh, part plastic and metal? Etchison asks: "What was the point — 25 percent 50 percent 75 percent? — at which the part that is human dies and the oiled, punched, guaranteed factory parts take over? How do you know when you stop being a person who is part machine, and become merely a machine that is part human?" The most logical step in prosthetics would obviously be the installation of a miniature computer in a human cranium to serve as the command and control center for artificial intelligence.

Such a step in the mechanization process leads us unequivocally to the case of the android, a subject whose wider implications may be discerned by a perusal of contemporary science fiction. In Norbert Wiener's *The Human Use of Human Beings* (1950), the author spoke of the Faustian compulsion to apply "know-how" without determining "know-what." This seems to be true of us Americans, who, unlike the Greeks, seldom view the harnessing of nature's powers with split emotions. Wiener speaks of the great figure of Prometheus, defying the gods of Olympus by bringing fire down from heaven to earth, as "the prototype of the scientist." If science constructs a thinking machine without any power of learning, then it will necessarily be a literal-minded servant of man. However, that is not, as we shall see, the ultimate development in bio-engineering. Here science fiction can aid by supplying a good number of robot scenarios which permit us in the early stages of thinking machines to ponder the social and ethical implications of lens-eyed mannikins replete with synthetic skin, steel tendons, ball-bearing joints and preset responses. With this in mind, we may now pass to a selective sampling of android fiction in order to pierce the curtain of the future.

One distinguished science fiction writer, Isaac Asimov, implies that programming circuits in servomechanisms could be made to follow the three master principles of the world's ethical systems: (1) Every robot is to protect men, even allowing himself to be destroyed if necessary; (2) Every robot must defer to legitimate authority, even if disagreeable; and (3) Every robot is to insure his own self-preservation as long as it is compatible with Rule 1. Asimov's Rules of Robotics are common enough in the plots of other authors so that detection of an android may pivot on some test involving a violation of the altruistic nature of robo-psychology.

Contrasted with the benevolent, albeit deterministic, behavior of androids is the range of purposes to which their human masters put them. Nowhere is the application of automation for automation's sake better seen than in the typical entertainment story involving androids. Often the customary behavior of androids serves to increase the convenience, the efficiency and the pleasures of human beings. One boxing manager trains his fighter assiduously for a championship bout by using an android as a sparring partner, an android capable of simulating the pugilistic style of any boxer,

living or dead. One night the fighter, without his manager's knowledge, inadvertently inserts in his mechanical "sparring partner" a programming subroutine which simulates himself. The result is an ironic death caused by fighting his automated *alter ego*. Richard Mathieson's story, *Steel*, deals with a similar Faustian compulsion to match human energy against mechanical force. Thus we find that "thinking robots," as highly intelligent and powerful means, are subservient to indifferent, at times trivial, ends.

In James Causey's *The Show Must Go On*, we see hate bars employed in order to siphon off the brimming aggressions of human beings. A similar therapy consists in the vicarious and cathartic sensations which human spectators at a circus derive in watching android performers accomplish perilous feats of strength and dexterity. The story ends with the accidental destruction of a female android by a beast. This pleases the crowd so much that every evening a similar tragedy is scheduled, thus recalling the "pan and circenses" psychology of the Roman amphitheatre in the days of Nero. Robots, as we see, are expendable in themselves, and must serve even such trivial ends of human beings as victims of a spectator sport.

Ron Goulart presents a less gruesome but no less disturbing story of android behavior in *Badinage*, a 1984-ish tale of the broad sanctions which a Credit Authority possesses in a totally cashless society. Goulart's android is a mechanical Iago, plotting to keep in perpetual indebtedness the story's protagonist, who, like Othello, finally recognizes the moral tilt in the robot's behavior and seeks to destroy him. Goulart has a subtle critical sense about the "law of irresistible use" implicit in the mechanization process, a law which seems to say that if a process can be automated, it will be automated. His imaginative concern has led to stories about android-staffed colleges, hospitals, old age homes and missions. Here the purposes which androids serve are more utilitarian, less casual.

The question of deviant behavior, implicit

in *Badinage,* is more elaborately developed in Shelley Lowenkopf's *The Addict.* In addition to a pathological attraction for devouring fiction literature, Lowenkopf's female robot has been programmed to feel shame and modesty as well as superiority in the presence of non-humanoid robots. (The implication of caste and racial discrimination is clear.) Lowenkopf suggests that the solution for eccentric androids is much simpler than for human beings — reprogramming of the computer tapes. This "robotherapy" eliminates anomic responses and guarantees social conformity according to the prevailing cultural "definition of the situation" in keeping with whatever is acceptable behavior. In these instances we have clearly medical and social ends being called into play regarding the disposition of automata.

The mechanization process not only permits progressive improvements on one and the same android but most especially between one generation of androids and a later one. Since robots may be improved till they acquire human appearance, the advanced model of android is capable of multiple responses and branching decisions as complex as those made by today's "man-in-the-street." In his story, *Evidence,* Isaac Asimov presents a superior android with a highly developed positronic brain. Having already served as district attorney, the android runs for the office of mayor and wins. Although neither omniscient nor infallible, the android mayor offers certain advantages to his constituency. Because of the first of the Three Laws of Robotics, he will never hurt human beings, even through corruption or tyranny. Undoubtedly the 21st century will see political decision-making being done by thinking robots.

Once androids reach a stage of technical development where they cannot be distinguished from human beings, the possibility arises of serving as convenient "stand-ins" for persons. The issue of "replica-presence," as it is called, furnishes material for a large number of "android" science writings, particularly in the ability of life-size marionettes to satisfy adequately emotional needs for company, conversation, and even love. In *The Changeling,* Ray Bradbury presents a wealthy man who, tired of his mistresses and desirous of solitude, purchases automatic replicas of himself to keep six women happy. Bradbury, unaware of the giant pun he created in this story of "sex-location," caught the essential preoccupation of those who, like T. S. Eliot, S. Giedion, Marshall McLuhan, Arnold Toynbee, Lancelot Law Whyte and Archibald MacLeish, lament the dissociation of thought and feeling in contemporary industrial society.

This split is described even more dramatically by Robert F. Young in *Juke Doll.* In this short story, Young depicts how the human void for affection is filled by a "doll-friend," who, even more than the paper-doll of popular music fame, is faithful to its human master. Speaking demurely in an electronic voice, this carefully assembled puppet is found to be merely the extension of the master-puppeteer's strategy. While all the other plots mentioned earlier have philosophical and ethical overtones, the case of the "humanoid" and its power of "replica-presence" presents moral and metaphysical difficulties of the first magnitude.

When does an android cease being a literal-minded slave of human masters and become autonomous, or truly humanoid? Norbert Wiener gives the answer in *The Human Use of Human Beings,* saying that the machine "which can learn and can make decisions on the basis of its learning, will in no way be obliged to make such decisions as we should have made, or will be acceptable to us." Such is the case in Henry Kuttner's *Those Among Us,* an attention-arresting plot about an android, cast so perfectly in the image of man as to represent a threat to his fellow androids. Situated in the twilight zone of the organic and the inorganic, this humanoid undergoes the identity crisis of "the stranger," that outsider who must traffic convivially with others who know exclusively those rules of the game inside of which they are meaningful actors. Kuttner taxes to the utmost the principle of identity which underlies Western rationalism from the time of Aristotle: Can A be A and non-A at one and the same time? In other words, can a machine be human? One of Kuttner's characters indicates the alarming convergence of a possible man-machine incarnation: "When the first successful humanoid was produced by the robots." Here we have a giant hall of mirrors with an infinite series of human and mechanical reflections and counter-reflections.

The same identity crisis emerges from William F. Nolan's *The Joy of Living*, in which a mechanical bride (apologies to Marshall McLuhan!) is accepted by a widower for both his personal solace and his children's tutelage. Having developed an affective tie for her human husband, the pseudo-wife, Margaret, pleads eloquently not to be discarded on the argument that, in creating her, man imparted his own qualities of compassion and feeling. One finishes the story with an ambivalent reaction, of deep awe and grateful acquiescence, not at all dissimilar to the mingled feeling inspired by the Genesis account of the creation of man from the slime of the earth and of woman from his side. We are at the heart of the dilemma posed by the mechanization process. Is not the technologist somewhat like Prometheus defying the limits placed by the Gods of Olympus, in order to bring fire from heaven to earth?

It is true, as Buckminster Fuller has pointed out, that the discoverer always ventures into the outlaw area. It is important, however, that we know the antinomies implicit in all frontier-transcending enterprises, the sense of awe which should accompany our decision to chart the unknown. Norbert Wiener says that the application of "know-how" with little "know-what" is definitely Promethean, inevitably bringing its own punishment with it. This typical awareness of the Greek outlook is needed if man is to refrain from the Promethean urge to transfer to the machine made in his likeness, the responsibility for his choice of good and evil, without continuing to accept a full responsibility for that choice. Therein should be the indissolubly wedded elements of awe and acquiescence, the "Allelujah" of creaturely respect with the "Amen" of human creation.

Mankind must awake to the discovery that we set an irreversible course. The writings of Pitt, McCullough, Wiener, Shannon, Pierce and Roseblith, to name a few of the leading names in the field of engineering biophysics, point to a cybernetic world where "feedback" arrangements could produce a high order of consciousness with a corresponding capacity for language and interpersonal communication. Is human voluntary action, after all, more than a choice among psychic tropisms so that the direction of commitment is given by the most useful resolution of decision vectors? If it is not, then androids will compete with men and perhaps overthrow them as Samuel Butler warned in the eighteenth century [*sic*]. His famous book, *Erewhon*, speaks of machines which could conquer mankind through the use of men as instrumental accomplices. If, on the other hand, man has an immortal soul which only God can create, then there can be no serious identity crisis as raised by mechanization's blind pursuit of technical virtuosity and efficiency. The crux is in the matter of ends and not just means. If our purposes are clear then we can hallow the means.

Philosophy and theology aside, the problem still remains for all of us, just as for scientists such as Norbert Wiener, of approximating human rational performance by machines. After all, the physical identity of an individual does not consist in matter alone but "in a certain continuity of process, and in the memory by the organism of the effects of its past development." This simulation of essentially spiritual characteristics is the basis of a very unusual story by Charles Beaumont called *Last Rites*. Extremely sick, a long-standing friend of a priest calls him to his bedside and raises the issue of a humanoid soul: "If your friend were suddenly to reveal himself to you as a machine, and he was dying, and wanted very much to go to Heaven — what would you do?" The suspense of the tale lies in whether the question is hypothetical or biographical. Is the priest's friend an authentic human or the faultless blueprint projection of some master bioengineer? Beaumont never tells us what the priest discovers after he anoints his friend.

One could go on documenting the insights raised by science fiction writers concerning "thinking robots." More imperative at this point is a public awareness that this "revolutionary revolution" to use Aldous Huxley's apt expression, is upon us. Before his death in California, Huxley told a reporter that his "Brave New World" was coming into being faster than he had anticipated. It is not yet a quarter of a century since Mark I was developed by Harvard's Howard Aiken. Since that time we have vastly improved on this cathode-ray tube model with miniaturized transistor models capable of playing chess, composing poetry and music, and designing art patterns. The idea of a tiny computer contained within an android skull is a logical step within the dynamics of the mechanization process. William F. Nolan says in the preface to his anthology, The Pseudo-People: "The birth of the first android, therefore, is a lot closer to us than we might imagine. Artificial hearts, lungs and arteries are already being developed in science; the artificial brain is the next major step toward the creation of humanized robots."

Ethicians, philosophers, moral theologians and behavioral scientists would do well to take science fiction more seriously, and most especially that body of literature which deals with robots, androids and humanoids. It is more possible to direct the mechanization process in the earlier stages of its course than when it has almost come to term in its definitive impact on society. Much concern over the spurious, even noxious, effects of technology arise out of the *fait accompli*, the distasteful, perhaps poisonous, fruit of seeds planted much earlier. In a very real sense, the future has already begun, as Robert Jungk wrote some years back. Those who may doubt that the future of androids has already begun should visit Disneyland in Southern California where regularly scheduled performances are held of a "replica-presence" of Abraham Lincoln: this truly amazing robot sits, stands, takes steps, makes gestures and speaks, even if imperfectly. If we can reproduce an Abraham Lincoln in Disneyland, what is to prevent science from bringing a Winston Churchill back to rule England?

LEFT: "Dorothy Wound Up Number One." The Tin Woodman who lacked a heart is undoubtedly the most celebrated mechanical man in all children's literature. But in the twelfth of L. Frank Baum's "Oz" books, we discover that he had, originally, been a flesh-and-blood woodman. A truly "authentic" mechanical man is the rotund Tik-Tok whom Dorothy first encounters in Ozma of Oz (1907). Eventually, the tubby robot gets a book to himself: Tik-Tok of Oz (1914), in which a new heroine, Betsy Bobbin, and her companion, the Shaggy Man, rescue Tik-Tok from a deep dark well. (Private Collection)

BELOW: Roly-poly robots were the speciality of such artists as Leo Morey, Julian S. Krupa, and J. E. Kelleam, who illustrated many of the stories in the science-fiction pulp magazines of the 1930s and 1940s. It is impossible to believe that their conceptions were not influenced by childhood memories of Tik-Tok of Oz. Here is an illustration by Krupa to Arthur R. Tofte's "Revolt of the Robots" in the May 1939 Fantastic Adventures. (New York Public Library, Picture Collection)

"She smoothes her hair with automatic hand,
And puts a record on the gramophone. . . ."
—T. S. Eliot, *"The Waste Land"*
(Front cover illustration by David Pelham for
Final Stage: The Ultimate Science Fiction
Anthology, *Penguin Books, 1975)*

Moxon's Master

by Ambrose Bierce

Bierce's Moxon is no Maelzel: his automatic chess-player is the real thing. Too real for its inventor's good. For when a machine begins to think like a man, who will ultimately control whom? Fortunately, the chess-player in this classic tale of the macabre, like the chess-playing, deadly computer in 2001: A *Space Odyssey*, belongs to the realm of fiction. But as our thinking machines become more and more complex, perhaps we should have less and less reason for remaining complacent about our mastery of them . . .

"Are you serious? — do you really believe that a machine thinks?"

I got no immediate reply; Moxon was apparently intent upon the coals in the grate, touching them deftly here and there with the fire-poker till they signified a sense of his attention by a brighter glow. For several weeks I had been observing in him a growing habit of delay in answering even the most trivial of commonplace questions. His air, however, was that of preoccupation rather than deliberation: one might have said that he had "something on his mind."

Presently he said:

"What is a 'machine'? The word has been variously defined. Here is one definition from a popular dictionary: 'Any instrument or organization by which power is applied and made effective, or a desired effect produced.' Well, then, is not a man a machine? And you will admit that he thinks — or thinks he thinks."

"If you do not wish to answer my question," I said, rather testily, "why not say so? — all that you say is mere evasion. You know well enough that when I say 'machine' I do not mean a man, but something that man has made and controls."

"When it does not control him," he said, rising abruptly and looking out of a window, whence nothing was visible in the blackness of a stormy night. A moment later he turned about and with a smile said: "I beg your pardon; I had no thought of evasion. I considered the dictionary man's unconscious testimony suggestive and worth something in the discussion. I can give your question a direct answer easily enough: I do believe that a machine thinks about the work that it is doing."

That was direct enough, certainly. It was not altogether pleasing, for it tended to confirm a sad suspicion that Moxon's devotion to study and work in his machine-shop had not been good for him. I knew, for one thing, that he suffered from insomnia, and that is no light affliction. Had it affected his mind? His reply to my question seemed to me then evidence that it had; perhaps I should think differently about it now. I was younger

From Ambrose Bierce, *Can Such Things Be?* (New York: Neale Publishing Co., 1903; second, enlarged edition). Reprinted by permission of Citadel Press.

then, and among the blessings that are not denied to youth is ignorance. Incited by that great stimulant to controversy, I said:

"And what, pray, does it think with — in the absence of a brain?"

The reply, coming with less than his customary delay, took his favorite form of counter-interrogation:

"With what does a plant think — in the absence of a brain?"

"Ah, plants also belong to the philosopher class! I should be pleased to know some of their conclusions; you may omit the premises."

"Perhaps," he replied, apparently unaffected by my foolish irony, "you may be able to infer their convictions from their acts. I will spare you the familiar examples of the sensitive mimosa, the several insectivorous flowers and those whose stamens bend down and shake their pollen upon the entering bee in order that he may fertilize their distant mates. But observe this. In an open spot in my garden I planted a climbing vine. When it was barely above the surface I set a stake into the soil a yard away. The vine at once made for it, but as it was about to reach it after several days I removed it a few feet. The vine at once altered its course, making an acute angle, and again made for the stake. This manœuvre was repeated several times, but finally, as if discouraged, the vine abandoned the pursuit and ignoring further attempts to divert it traveled to a small tree, further away, which it climbed.

"Roots of the eucalyptus will prolong themselves incredibly in search of moisture. A well-known horticulturist relates that one entered an old drain pipe and followed it until it came to a break, where a section of the pipe had been removed to make way for a stone wall that had been built across its course. The root left the drain and followed the wall until it found an opening where a stone had fallen out. It crept through and following the other side of the wall back to the drain, entered the unexplored part and resumed its journey."

"And all this?"

"Can you miss the significance of it? It shows the consciousness of plants. It proves that they think."

"Even if it did — what then? We were speaking, not of plants, but of machines. They may be composed partly of wood — wood that has no longer vitality — or wholly of metal. Is thought an attribute also of the mineral kingdom?"

"How else do you explain the phenomena, for example, of crystallization?"

"I do not explain them."

"Because you cannot without affirming what you wish to deny, namely, intelligent co-operation among the constituent elements of the crystals. When soldiers form lines, or hollow squares, you call it reason. When wild geese in flight take the form of a letter V you say instinct. When the homogeneous atoms of a mineral, moving freely in solution, arrange themselves into shapes mathematically perfect, or particles of frozen moisture into the symmetrical and beautiful forms of snowflakes, you have nothing to say. You have not even invented a name to conceal your heroic unreason."

Moxon was speaking with unusual animation and earnestness. As he paused I heard in an adjoining room known to me as his "machine-shop," which no one but himself was permitted to enter, a singular thumping sound, as of some one pounding upon a table with an open hand. Moxon heard it at the same moment and, visibly agitated, rose and hurriedly passed into the room whence it came. I thought it odd that any one else should be in there, and my interest in my friend — with doubtless a touch of unwarrantable curiosity — led me to listen intently, though, I am happy to say, not at the keyhole. There were confused sounds, as of a struggle or scuffle; the floor shook. I distinctly heard hard breathing and a hoarse whisper which said "Damn you!" Then all was silent, and presently Moxon reappeared and said, with a rather sorry smile:

"Pardon me for leaving you so abruptly.

I have a machine in there that lost its temper and cut up rough."

Fixing my eyes steadily upon his left cheek, which was traversed by four parallel excoriations showing blood, I said:

"How would it do to trim its nails?"

I could have spared myself the jest; he gave it no attention, but seated himself in the chair that he had left and resumed the interrupted monologue as if nothing had occurred:

"Doubtless you do not hold with those (I need not name them to a man of your reading) who have taught that all matter is sentient, that every atom is a living, feeling, conscious being. I do. There is no such thing as dead, inert matter: it is all alive; all instinct with force, actual and potential; all sensitive to the same forces in its environment and susceptible to the contagion of higher and subtler ones residing in such superior organisms as it may be brought into relation with, as those of man when he is fashioning it into an instrument of his will. It absorbs something of his intelligence and purpose — more of them in proportion to the complexity of the resulting machine and that of its work.

"Do you happen to recall Herbert Spencer's definition of 'Life'? I read it thirty years ago. He may have altered it afterward, for anything I know, but in all that time I have been unable to think of a single word that could profitably be changed or added or removed. It seems to me not only the best definition, but the only possible one.

" 'Life,' he says, 'is a definite combination of heterogeneous changes, both simultaneous and successive, in correspondence with external co-existences and sequences.' "

"That defines the phenomenon," I said, "but gives no hint of its cause."

"That," he replied, "is all that any definition can do. As Mill points out, we know nothing of cause except as an antecedent — nothing of effect except as a consequent. Of certain phenomena, one never occurs without another, which is dissimilar: the first in point of time we call cause, the second,

effect. One who had many times seen a rabbit pursued by a dog, and had never seen rabbits and dogs otherwise, would think the rabbit the cause of the dog.

"But I fear," he added, laughing naturally enough, "that my rabbit is leading me a long way from the track of my legitimate quarry: I'm indulging in the pleasure of the chase for its own sake. What I want you to observe is that in Herbert Spencer's definition of 'life' the activity of a machine is included — there is nothing in the definition that is not applicable to it. According to this sharpest of observers and deepest of thinkers, if a man during his period of activity is alive, so is a machine when in operation. As an inventor and constructor of machines I know that to be true."

Moxon was silent for a long time, gazing absently into the fire. It was growing late and I thought it time to be going, but somehow I did not like the notion of leaving him in that isolated house, all alone except for the presence of some person of whose nature my conjectures could go no further than that it was unfriendly, perhaps malign. Leaning toward him and looking earnestly into his eyes while making a motion with my hand through the door of his workshop, I said:

"Moxon, whom have you in there?"

Somewhat to my surprise he laughed lightly and answered without hesitation:

"Nobody; the incident that you have in mind was caused by my folly in leaving a machine in action with nothing to act upon, while I undertook the interminable task of enlightening your understanding. Do you happen to know that Consciousness is the creature of Rhythm?"

"O bother them both!" I replied, rising and laying hold of my overcoat. "I'm going to wish you good night; and I'll add the hope that the machine which you inadvertently left in action will have her gloves on the next time you think it needful to stop her."

Without waiting to observe the effect of my shot I left the house.

Rain was falling, and the darkness was

intense. In the sky beyond the crest of a hill toward which I groped my way along precarious plank sidewalks and across miry, unpaved streets I could see the faint glow of the city's lights, but behind me nothing was visible but a single window of Moxon's house. It glowed with what seemed to me a mysterious and fateful meaning. I knew it was an uncurtained aperture in my friend's "machine-shop," and I had little doubt that he had resumed the studies interrupted by his duties as my instructor in mechanical consciousness and the fatherhood of Rhythm. Odd, and in some degree humorous, as his convictions seemed to me at that time, I could not wholly divest myself of the feeling that they had some tragic relation to his life and character — perhaps to his destiny — although I no longer entertained the notion that they were the vagaries of a disordered mind. Whatever might be thought of his views, his exposition of them was too logical for that. Over and over, his last words came back to me: "Consciousness is the creature of Rhythm." Bald and terse as the statement was, I now found it infinitely alluring. At each recurrence it broadened in meaning and deepened in suggestion. Why, here, (I thought) is something upon which to found a philosophy. If consciousness is the product of rhythm all things *are* conscious, for all have motion, and all motion is rhythmic. I wondered if Moxon knew the significance and breadth of his thought — the scope of this momentous generalization; or had he arrived at his philosophic faith by the tortuous and uncertain road of observation?

That faith was then new to me, and all Moxon's expounding had failed to make me a convert; but now it seemed as if a great light shone about me, like that which fell upon Saul of Tarsus; and out there in the storm and darkness and solitude I experienced what Lewes calls "the endless variety and excitement of philosophic thought." I exulted in a new sense of knowledge, a new pride of reason. My feet seemed hardly to touch the earth; it was as if I were uplifted and borne through the air by invisible wings.

Yielding to an impulse to seek further light from him whom I now recognized as my master and guide, I had unconsciously turned about, and almost before I was aware of having done so found myself again at Moxon's door. I was drenched with rain, but felt no discomfort. Unable in my excitement to find the doorbell I instinctively tried the knob. It turned and, entering, I mounted the stairs to the room that I had so recently left. All was dark and silent; Moxon, as I had supposed, was in the adjoining room — the "machine-shop." Groping along the wall until I found the communicating door I knocked loudly several times, but got no response, which I attributed to the uproar outside, for the wind was blowing a gale and dashing the rain against the thin walls in sheets. The drumming upon the shingle roof spanning the unceiled room was loud and incessant.

I had never been invited into the machine-shop — had, indeed, been denied admittance, as had all others, with one exception, a skilled metal worker, of whom no one knew anything except that his name was Haley and his habit silence. But in my spiritual exaltation, discretion and civility were alike forgotten and I opened the door. What I saw took all philosophical speculation out of me in short order.

Moxon sat facing me at the farther side of a small table upon which a single candle made all the light that was in the room. Opposite him, his back toward me, sat another person. On the table between the two was a chess-board; the men were playing. I knew little of chess, but as only a few pieces were on the board it was obvious that the game was near its close. Moxon was intensely interested — not so much, it seemed to me, in the game as in his antagonist, upon whom he had fixed so intent a look that, standing though I did directly in the line of his vision, I was altogether unobserved. His face was ghastly white, and his eyes glittered like diamonds. Of his antagonist I had only a back view, but that was sufficient; I should not have cared to see his face.

He was apparently not more than five feet in height, with proportions suggesting those of a gorilla — a tremendous breadth of shoulders, thick, short neck and broad, squat head, which had a tangled growth of black hair and was topped with a crimson fez. A tunic of the same color, belted tightly to the waist, reached the seat — apparently a box — upon which he sat; his legs and feet were not seen. His left forearm appeared to rest in his lap; he moved his pieces with his right hand, which seemed disproportionately long.

I had shrunk back and now stood a little to one side of the doorway and in shadow. If Moxon had looked farther than the face of his opponent he could have observed nothing now, except that the door was open. Something forbade me either to enter or to retire, a feeling — I know not how it came — that I was in the presence of an imminent tragedy and might serve my friend by remaining. With a scarcely conscious rebellion against the indelicacy of the act I remained.

The play was rapid. Moxon hardly glanced at the board before making his moves, and to my unskilled eye seemed to move the piece most convenient to his hand, his motions in doing so being quick, nervous and lacking in precision. The response of his antagonist, while equally prompt in the inception, was made with a slow, uniform, mechanical and, I thought, somewhat theatrical movement of the arm, that was a sore trial to my patience. There was something unearthly about it all, and I caught myself shuddering. But I was wet and cold.

Two or three times after moving a piece the stranger slightly inclined his head, and each time I observed that Moxon shifted his king. All at once the thought came to me that the man was dumb. And then that he was a machine — an automaton chess-player! Then I remembered that Moxon had once spoken to me of having invented such a piece of mechanism, though I did not un-

derstand that it had actually been constructed. Was all his talk about the consciousness and intelligence of machines merely a prelude to eventual exhibition of this device — only a trick to intensify the effect of its mechanical action upon me in my ignorance of its secret?

A fine end, this, of all my intellectual transports — my "endless variety and excitement of philosophic thought!" I was about to retire in disgust when something occurred to hold my curiosity. I observed a shrug of the thing's great shoulders, as if it were irritated: and so natural was this — so entirely human — that in my new view of the matter it startled me. Nor was that all, for a moment later it struck the table sharply with its clenched hand. At that gesture Moxon seemed even more startled than I: he pushed his chair a little backward, as in alarm.

Presently Moxon, whose play it was, raised his hand high above the board, pounced upon one of his pieces like a sparrow-hawk and with the exclamation "checkmate!" rose quickly to his feet and stepped behind his chair. The automaton sat motionless.

The wind had now gone down, but I heard, at lessening intervals and progressively louder, the rumble and roll of thunder. In the pauses between I now became conscious of a low humming or buzzing which, like the thunder, grew momentarily louder and more distinct. It seemed to come from the body of the automaton, and was unmistakably a whirring of wheels. It gave me the impression of a disordered mechanism which had escaped the repressive and regulating action of some controlling part — an effect such as might be expected if a pawl should be jostled from the teeth of a ratchet-wheel. But before I had time for much conjecture as to its nature my attention was taken by the strange motions of the automation itself. A slight but continuous convulsion appeared to have possession of it. In body and head it shook like a man with palsy or an ague chill, and the motion augmented every moment until the entire figure was in violent agitation. Suddenly it sprang to its feet and with a movement almost too quick for the eye to follow shot forward across table and chair, with both arms thrust forth to their full length — the posture and lunge of a diver. Moxon tried to throw himself backward out of reach, but he was too late: I saw the horrible thing's hands close upon his throat, his own clutch its wrists. Then the table was overturned, the candle thrown to the floor and extinguished, and all was black dark. But the noise of the struggle was dreadfully distinct, and most terrible of all were the raucous, squawking sounds made by the strangled man's efforts to breathe. Guided by the infernal hubbub, I sprang to the rescue of my friend, but had hardly taken a stride in the darkness when the whole room blazed with a blinding white light that burned into my brain and heart and memory a vivid picture of the combatants on the floor, Moxon underneath, his throat still in the clutch of those iron hands, his head forced backward, his eyes protruding, his mouth wide open and his tongue thrust out; and — horrible contrast! — upon the painted face of his assassin an expression of tranquil and profound thought, as in the solution of a problem in chess! This I observed, then all was blackness and silence.

Three days later I recovered consciousness in a hospital. As the memory of that tragic night slowly evolved in my ailing brain I recognized in my attendant Moxon's confidential workman, Haley. Responding to a look he approached, smiling.

"Tell me about it," I managed to say, faintly — "all about it."

"Certainly," he said; "you were carried unconscious from a burning house — Moxon's. Nobody knows how you came to be there. You may have to do a little explaining. The origin of the fire is a bit mysterious, too. My own notion is that the house was struck by lightning."

"And Moxon?"

"Buried yesterday — what was left of him."

Apparently this reticent person could unfold himself on occasion. When imparting shocking intelligence to the sick he was affable enough. After some moments of the keenest mental suffering I ventured to ask another question:

"Who rescued me?"

"Well, if that interests you — I did."

"Thank you, Mr. Haley, and may God bless you for it. Did you rescue, also, that charming product of your skill, the automaton chess-player that murdered its inventor?"

The man was silent a long time, looking away from me. Presently he turned and gravely said:

"Do you know that?"

"I do," I replied; "I saw it done."

That was many years ago. If asked to-day I should answer less confidently.

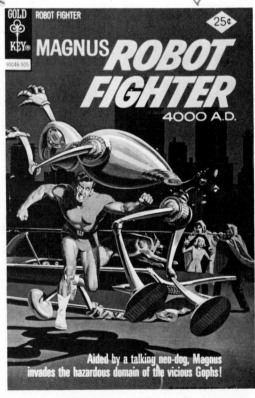

ABOVE:
*(Copyright © 1978 by Marvel Comics Group.
All rights reserved.)*

RIGHT:
(From MAGNUS ROBOT FIGHTER 4000 A.D. *No.
16, © 1966 Western Publishing Company,
Inc. Reprinted by permission of the publisher.)*

(Courtesy Manuel Rodriguez)

Helen O'Loy

by Lester del Rey

Lester del Rey's story (originally published in *Astounding*, December 1938), together with Isaac Asimov's "I, Robot" (originally published in *Amazing*, January
1939), radically changed science fiction attitudes to robots. Damon Knight (*In
Search of Wonder*, 1967, p. 133) quotes a relevant but undocumented comment
by Howard Browne: "[S]uch machines were invariably depicted as potential Frankenstein monsters, as trustworthy as a tiger and apt to turn upon their creators at any
moment with or without provocation. This approach became a tiresome pattern
in science fiction. . . ." Under the influence of Del Rey and Asimov, science fiction
robots now became sensitive and trustworthy; they had more to fear from unpredictable human beings than vice versa, and their former, less amiable characteristics
became the attributes of androids.

I am an old man now, but I can still see
Helen as Dave unpacked her, and still hear
him gasp as he looked her over.

"Man, isn't she a beauty?"

She was beautiful, a dream in spun plas-
tics and metals, something Keats might have
seen dimly when he wrote his sonnet. If
Helen of Troy had looked like that, the
Greeks must have been pikers when they
launched only a thousand ships; at least,
that's what I told Dave.

"Helen of Troy, eh?" He looked at her tag.
"At least it beats this thing — K2W88.
Helen . . . hm-m-m . . . Helen of Alloy."

"Not much swing to that, Dave. Too
many unstressed syllables in the middle.
How about Helen O'Loy?"

"Helen O'Loy she is, Phil." And that's
how it began — one part beauty, one part
dream, one part science; add a stereo broad-

cast, stir mechanically, and the result is
chaos.

Dave and I hadn't gone to college together, but when I came to Messina to practice
medicine, I found him downstairs in a little
robot repair shop. After that, we began to
pal around, and when I started going with
one twin, he found the other equally attractive, so we made it a foursome.

When our business grew better, we rented
a house out near the rocket field — noisy
but cheap, and the rockets discouraged apartment building. We liked room enough to
stretch ourselves. I suppose if we hadn't
quarreled with them, we'd have married the
twins in time. But Dave wanted to look
over the latest Venus-rocket attempt when
his twin wanted to see a display stereo
starring Larry Ainslee, and they were both
stubborn. From then on, we forgot the
girls and spent our evenings at home.

But it wasn't until "Lena" put vanilla
on our steak instead of salt that we got off
on the subject of emotions and robots.

While Dave was dissecting Lena to find the trouble, we naturally mulled over the future of the mechs. He was sure that the robots would beat men some day, and I couldn't see it.

"Look here, Dave," I argued. "You know Lena doesn't think — not really. When those wires crossed, she could have corrected herself. But she didn't bother; she followed the mechanical impulse. A man might have reached for the vanilla, but when he saw it in his hand, he'd have stopped. Lena has sense enough, but she has no emotions, no consciousness of self."

"All right, that's the big trouble with the mechs now. But we'll get around it, put in some mechanical emotions, or something." He screwed Lena's head back on, turned on her juice. "Go back to work, Lena. It's nineteen o'clock."

Now I specialized in endocrinology and related subjects. I wasn't exactly a psychologist, but I did understand the glands, secretions, hormones, and miscellanies that are the physical causes of emotions. It took medical science three hundred years to find out how and why they worked, and I couldn't see men duplicating them mechanically in much less time.

I brought home books and papers to prove it, and Dave quoted the invention of memory coils and veritoid eyes. During that year we swapped knowledge until Dave knew the whole theory of endocrinology, and I could have made Lena from memory. The more we talked, the less sure I grew about the impossibility of *homo mechanensis* as the perfect type.

Poor Lena. Her cuproberyl body spent half its time in scattered pieces. Our first attempts were successful only in getting her to serve fried brushes for breakfast and wash the dishes in oleo oil. Then one day she cooked a perfect dinner with six wires crossed, and Dave was in ecstasy.

He worked all night on her wiring, put in a new coil, and taught her a fresh set of words. And the next day she flew in a tantrum and swore vigorously at us when we told her she wasn't doing her work right.

"It's a lie," she yelled, shaking a suction brush. "You're all liars. If you so-and-so's would leave me whole long enough, I might get something done around the place."

When we calmed her temper and got her back to work, Dave ushered me into the study. "Not taking any chances with Lena," he explained. "We'll have to cut out that adrenal pack and restore her to normalcy. But we've got to get a better robot. A house-maid mech isn't complex enough."

"How about Dillard's new utility models? They seem to combine everything in one."

"Exactly. Even so, we'll need a special one built to order, with a full range of memory coils. And out of respect to old Lena, let's get a female case for the works."

The result, of course, was Helen. The Dillard people had performed a miracle and put all the works in a girl-modeled case. Even the plastic and rubberite face was designed for flexibility to express emotions, and she was complete with tear glands and taste buds, ready to simulate every human action from breathing to pulling hair. The bill they sent with her was another miracle but Dave and I scraped it together; we had to turn Lena over to an exchange to complete it, though, and thereafter we ate out.

I'd performed plenty of delicate operations on living tissues, and some of them had been tricky, but I still felt like a pre-med student as we opened the front plate of her torso and began to sever the leads of her "nerves." Dave's mechanical glands were all prepared, complex little bundles of radio tubes and wires that heterodyned on the electrical thought impulses and distorted them as adrenalin distorts the reaction of human minds.

Instead of sleeping that night, we pored over the schematic diagrams of her structure, tracing the thought mazes of her wiring,

217

serving the leaders, implanting the heter-ones, as Dave called them. And while we worked a mechanical tape fed carefully pre-pared thoughts of consciousness and aware-ness of life and feeling into an auxiliary memory coil. Dave believed in leaving nothing to chance.

It was growing light as we finished, ex-hausted but exultant. All that remained was the starting of her electrical power; like all the Dillard mechs, she was equipped with a tiny atomotor instead of batteries, and once started would need no further atten-tion.

Dave refused to turn her on. "Wait until we've slept and rested," he advised. "I'm as eager to try her as you are, but we can't do much studying with our minds half dead. Turn in, and we'll leave Helen until later."

Even though we were both reluctant to follow it, we knew the idea was sound. We turned in, and sleep hit us before the air-conditioner could cut down to sleeping temperature. And then Dave was pound-ing on my shoulder.

"Phil! Hey, snap out of it!"

I groaned, turned over, and faced him. "Well? . . . Uh! What is it? Did Helen—"

"No, it's old Mrs. van Styler. She 'visored to say her son has an infatuation for a servant girl, and wants you to come out and give counter-hormones. They're at the summer camp in Maine."

Rich Mrs. van Styler! I couldn't afford to let that account down, now that Helen had used up the last of my funds. But it wasn't a job I cared for.

"Counter-hormones! That'll take two weeks' full time. Anyway, I'm no society doctor, messing with glands to keep fools happy. My job's taking care of serious trouble."

"And you want to watch Helen." Dave was grinning, but he was serious, too. "I told her it'd cost her fifty thousand."

"Huh?"

"And she said O.K., if you hurried."

Of course there was only one thing to do, though I could have wrung fat Mrs. van Styler's neck cheerfully. It wouldn't have happened if she'd used robots like everyone else — but she had to be different.

Consequently, while Dave was back home puttering with Helen, I was racking my brain to trick Archy van Styler into getting the counter-hormones, and giving the serv-ant girl the same. Oh, I wasn't supposed to, but the poor kid was crazy about Archy. Dave might have written, I thought, but never a word did I get.

It was three weeks later instead of two when I reported that Archy was "cured," and collected on the line. With that money in my pocket, I hired a personal rocket and was back in Messina in half an hour. I didn't waste time in reaching the house.

As I stepped into the alcove, I heard a light patter of feet, and an eager voice called out, "Dave, dear?" For a minute I couldn't answer, and the voice came again, pleading. "Dave?"

I don't know what I expected, but I didn't expect Helen to meet me that way, stopping and staring at me, obvious disap-pointment on her face, little hands fluttering up against her breast.

"Oh," she cried. "I thought it was Dave. He hardly comes home to eat now, but I've had supper waiting hours." She dropped her hands and managed a smile. "You're Phil, aren't you? Dave told me about you when . . . at first. I'm so glad to see you home, Phil."

"Glad to see you doing so well, Helen." Now what does one say for light conversa-tion with a robot? "You said something about supper?"

"Oh, yes. I guess Dave ate downtown again, so we might as well go in. It'll be nice having someone to talk to around the house, Phil. You don't mind if I call you Phil, do you? You know, you're sort of a godfather to me."

We ate. I hadn't counted on such behavior, but apparently she considered it as normal as walking. She didn't do much eating, at that; most of the time she spent staring at the front door.

Dave came in as we were finishing, a frown a yard wide on his face. Helen started to rise, but he ducked toward the stairs, throwing words over his shoulder.

"Hi, Phil. See you up here later."

There was something radically wrong with him. For a moment I'd thought his eyes were haunted, and as I turned to Helen, hers were filling with tears. She gulped, choked them back, and fell to viciously on her food.

"What's the matter with him . . . and you?" I asked.

"He's sick of me." She pushed her plate away and got up hastily. "You'd better see him while I clean up. And there's nothing wrong with me. And it's not my fault, anyway." She grabbed the dishes and ducked into the kitchen; I could have sworn she was crying.

Maybe all thought is a series of conditioned reflexes — but she certainly had picked up a lot of conditioning while I was gone. Lena in her heyday had been nothing like this. I went up to see if Dave could make any sense out of the hodgepodge.

He was squirting soda into a large glass of apple brandy, and I saw that the bottle was nearly empty. "Join me?" he asked.

It seemed like a good idea. The roaring blast of an ion rocket overhead was the only familiar thing left in the house. From the look around Dave's eyes, it wasn't the first bottle he'd emptied while I was gone, and there were more left. He dug out a new bottle for his own drink.

"Of course it's none of my business, Dave, but that stuff won't steady your nerves any. What's gotten into you and Helen? Been seeing ghosts?"

Helen was wrong; he hadn't been eating downtown — nor anywhere else. His muscles collapsed into a chair in a way that spoke of fatigue and nerves, but mostly of hunger. "You noticed it, eh?"

"Noticed it? The two of you jammed it down my throat."

"Uhm-m-m." He swatted at a non-existent fly, and slumped farther down in the pneumatic. "Guess maybe I should have waited with Helen until you got back. But if that stereo cast hadn't changed . . . anyway, it did. And those mushy books of yours finished the job."

"Thanks. That makes it all clear."

"You know, Phil, I've got a place up in the country . . . fruit ranch. My dad left it to me. Think I'll look it over."

And that's the way it went. But finally, by much liquor and more perspiration, I got some of the story out of him before I gave him an amytal and put him to bed. Then I hunted up Helen and dug the rest of the story from her until it made sense.

Apparently as soon as I was gone Dave had turned her on and made preliminary tests, which were entirely satisfactory. She had reacted beautifully — so well that he'd decided to leave her and go down to work as usual.

Naturally, with all her untried reactions, she was filled with curiosity, and wanted him to stay. Then he had an inspiration. After showing her what her duties about the house would be he set her down in front of the stereovisor, tuned in a travelogue, and left her to occupy her time with that.

The travelogue held her attention until it was finished, and the station switched on a current serial with Larry Ainslee, the same cute emoter who'd given us all the trouble with the twins. Incidentally, he looked something like Dave.

Helen took to the serial like a seal to water. This play acting was a perfect outlet for her newly excited emotions. When that particular episode finished she found a love story on another station, and added still more to her education. The afternoon programs were mostly news and music, but by

219

then she'd found my books; and I do have rather adolescent taste in literature.

Dave came home in the best of spirits. The front alcove was neatly swept, and there was the odor of food in the air that he'd missed around the house for weeks. He had visions of Helen as the super-efficient housekeeper.

So it was a shock to him to feel two strong arms around his neck from behind and hear a voice all aquiver coo into his ears, "Oh, Dave, darling, I've missed you so, and I'm so *thrilled* that you're back." Helen's technique may have lacked polish, but it had enthusiasm, as he found when he tried to stop her from kissing him. She had learned fast and furiously — also, Helen was powered by an atomotor.

Dave wasn't a prude, but he remembered that she was only a robot, after all. The fact that she felt, acted, and looked like a young goddess in his arms didn't mean much. With some effort, he untangled her and dragged her off to supper, where he made her eat with him to divert her attention.

After her evening work, he called her into the study and gave her a thorough lecture on the folly of her ways. It must have been good, for it lasted three solid hours, and covered her station in life, the idiocy of stereos, and various other miscellanies. When he finished, Helen looked up with dewy eyes and said wistfully, "I know, Dave, but I still love you."

That's when Dave started drinking.

It grew worse each day. If he stayed downtown, she was crying when he came home. If he returned on time, she fussed over him and threw herself at him. In his room, with the door locked, he could hear her downstairs pacing up and down and muttering; and when he went down, she stared at him reproachfully until he had to go back up.

I sent Helen out on a fake errand in the morning and got Dave up. With her gone, I made him eat a decent breakfast and gave him a tonic for his nerves. He was still listless and moody.

"Look here, Dave," I broke in on his brooding. "Helen isn't human, after all. Why not cut off her power and change a few memory coils? Then we can convince her that she never was in love and couldn't get that way."

"You try it. I had that idea, but she put up a wail that would wake Homer. She says it would be murder — and the hell of it is that I can't help feeling the same way about it. Maybe she isn't human, but you wouldn't guess it when she puts on that martyred look and tells you to go ahead and kill her."

"We never put in substitutes for some of the secretions present in man during the love period."

"I don't know what we put in. Maybe the heterones backfired or something. Anyway, she's made this idea so much a part of her thoughts that we'd have to put in a whole new set of coils."

"Well, why not?"

"Go ahead. You're the surgeon of the family. I'm not used to fussing with emotions. Matter of fact, since she's been acting this way, I'm beginning to hate work on any robot. My business is going to blazes."

He saw Helen coming up the walk and ducked out the back door for the monorail express. I'd intended to put him back in bed, but let him go. Maybe he'd be better off at his shop than at home.

"Dave's gone?" Helen did have that martyred look now.

"Yeah. I got him to eat, and he's gone to work."

"I'm glad he ate." She slumped down in a chair as if she were worn out, though how a mech could be tired beat me. "Phil?"

"Well, what is it?"

"Do you think I'm bad for him? I mean, do you think he'd be happier if I weren't here?"

"He'll go crazy if you keep acting this way around him."

She winced. Those little hands were twisting about pleadingly, and I felt like an inhuman brute. But I'd started, so I went ahead. "Even if I cut out your power and changed your coils, he'd probably still be haunted by you."

"I know. But I can't help it. And I'd make him a good wife, really I would, Phil."

I gulped; this was getting a little too far. "And give him strapping sons to boot, I suppose. A man wants flesh and blood, not rubber and metal."

"Don't, please! I can't think of myself that way; to me, I'm a woman. And you know how perfectly I'm made to imitate a real woman . . . in all ways. I couldn't give him sons, but in every other way . . . I'd try so hard, I know I'd make a good wife."

I gave up.

Dave didn't come home that night, nor the next day. Helen was fussing and fuming, wanting me to call the hospitals and the police, but I knew nothing had happened to him. He always carried identification. Still, when he didn't come in the third day, I began to worry. And when Helen started out for his shop, I agreed to go with her.

Dave was there, with another man I didn't know. I parked Helen where he couldn't see her, but where she could hear, and went in as soon as the other fellow left.

Dave looked a little better and seemed glad to see me. "Hi, Phil — just closing up. Let's go eat."

Helen couldn't hold back any longer, but came trooping in. "Come on home, Dave. I've got roast duck with spice stuffing, and you know you love that."

"Scat!" said Dave. She shrank back, turned to go. "Oh, all right, stay. You might as well hear it, too. I've sold the shop. The fellow you saw just bought it, and I'm going up to the old fruit ranch I told you about, Phil. I can't stand the mechs any more."

"You'll starve to death at that," I told him.

"No, there's a growing demand for old-fashioned fruit, raised out of doors. People are tired of this water-culture stuff. Dad always made a living out of it. I'm leaving as soon as I can get home and pack."

Helen clung to her idea. "I'll pack, Dave, while you eat. I've got apple cobbler for dessert." The world was toppling under her feet, but she still remembered how crazy he was for apple cobbler.

Helen was a good cook; in fact, she was a genius, with all the good points of a woman and a mech combined. Dave ate well enough, after he got started. By the time supper was over, he'd thawed out enough to admit he liked the duck and cobbler, and to thank her for packing. In fact, he even let her kiss him good-by, though he firmly refused to let her go to the rocket field with him.

Helen was trying to be brave when I got back, and we carried on a stumbling conversation about Mrs. van Styler's servants for a while. But the talk began to lull, and she sat staring out of the window at nothing most of the time. Even the stereo comedy lacked interest for her, and I was glad enough to have her go off to her room. She could cut her power down to simulate sleep when she chose.

As the days slipped by, I began to realize why she couldn't believe herself a robot. I got to thinking of her as a girl and companion myself. Except for odd intervals when she went off by herself to brood, or when she kept going to the telescript for a letter that never came, she was as good a companion as a man could ask. There was something homey about the place that Lena had never put there.

I took Helen on a shopping trip to Hudson and she giggled and purred over the wisps of silk and glassheen that were the fashion, tried on endless hats, and conducted herself as any normal girl might. We went trout fishing for a day, where she proved to be as good a sport and as sensibly silent as a man. I thoroughly enjoyed myself and thought she was forgetting Dave. That was before I came home unexpectedly and found her doubled up on the couch, threshing her legs up and down and crying to the high heavens.

It was then I called Dave. They seemed

221

to have trouble in reaching him and Helen came over beside me while I waited. She was tense and fidgety as an old maid trying to propose. But finally they located Dave.

"What's up, Phil?" he asked as his face came on the viewplate. "I was just getting my things together to — "

I broke him off. "Things can't go on the way they are, Dave. I've made up my mind. I'm yanking Helen's coils tonight. It won't be worse than what she's going through now."

Helen reached up and touched my shoulder. "Maybe that's best, Phil. I don't blame you."

Dave's voice cut in. "Phil, you don't know what you're doing!"

"Of course I do. It'll all be over by the time you can get here. As you heard, she's agreeing."

There was a black cloud sweeping over Dave's face. "I won't have it, Phil. She's half mine, and I forbid it!"

"Of all the — "

"Go ahead, call me anything you want. I've changed my mind. I was packing to come home when you called."

Helen jerked around me, her eyes glued to the panel. "Dave, do you . . . are you — "

"I'm just waking up to what a fool I've been, Helen. Phil, I'll be home in a couple of hours, so if there's anything — "

He didn't have to chase me out. But I heard Helen cooing something about loving to be a rancher's wife before I could shut the door.

Well, I wasn't as surprised as they thought. I think I knew when I called Dave what would happen. No man acts the way Dave had been acting because he hates a girl; only because he thinks he does — and thinks wrong.

No woman ever made a lovelier bride or sweeter wife. Helen never lost her flare for cooking and making a home. With her

gone, the old house seemed empty, and I began to drop out to the ranch once or twice a week. I suppose they had troubles at times, but I never saw it, and I know the neighbors never suspected they were anything but normal man and wife.

Dave grew older, and Helen didn't, of course. But between us, we put lines in her face and grayed her hair without letting Dave know that she wasn't growing old with him; he'd forgotten that she wasn't human, I guess.

I practically forgot, myself. It wasn't until a letter came from Helen this morning that I woke up to reality. There, in her beautiful script, just a trifle shaky in places, was the inevitable that neither Dave nor I had seen.

Dear Phil:

As you know, Dave has had heart trouble for several years now. We expected him to live on just the same, but it seems that wasn't to be. He died in my arms just before sunrise. He sent you his greetings and farewell.

I've one last favor to ask of you, Phil. There is only one thing for me to do when this is finished. Acid will burn out metal as well as flesh, and I'll be dead with Dave. Please see that we are buried together, and that the morticians do not find my secret. Dave wanted it that way, too.

Poor, dear Phil. I know you loved Dave as a brother, and how you felt about me. Please don't grieve too much for us, for we have had a happy life together, and both feel that we should cross this last bridge side by side.

With love and thanks from,

Helen

It had to come sooner or later, I suppose, and the first shock has worn off now. I'll be leaving in a few minutes to carry out Helen's last instructions.

Dave was a lucky man, and the best friend I ever had. And Helen — Well, as I've said, I'm an old man now, and can view things more sanely; I should have married and raised a family, I suppose. But . . . there was only one Helen O'Loy.

Percy Lund, drawing of a servomechanism, c. 1952. The eye in this bionic image of man-machine looks back toward an idea (for example, a signal to move a gun turret) and then looks below to see where the gun turret actually is. The eye then coordinates the difference between the two. (M.I.T. Historical Collections)

The Perfect Robot

by Pierre Boulle

The author of *Planet of the Apes* imaginatively explores the very questions discussed earlier by Professors Cohen and Ziff. What are the differences between human beings and automata? Can a robot ever be a person? M. Boulle's professor unexpectedly comes up with some different conclusions . . .

As Professor Fontaine came into the board room, the eyes of all the directors converged on the corrugated and voluminous cranium that housed the Company's most precious assets.

'Well?' the Board anxiously enquired.

A flame of triumph glittered in the old scientist's eyes.

'*Eureka*, gentlemen,' was all he said.

A tremor of relief went down the green baize table, and the prince of science was asked to report on his findings.

In its early days the EBC (Electronic Brain Company) used to manufacture modest calculating machines capable of carrying out simple arithmetical operations. The scope of these machines had been enormously enlarged, and their ingenuity had reached a degree of perfection unequalled by any rival concern, ever since the EBC had obtained the full-time services of Professor Fontaine, one of the most expert and enthusiastic pioneers of the young science of cybernetics.

After the directors had voted him a princely salary and installed him in a laboratory equipped with the latest research apparatus, the scientist, sheltered from all ma-

terial worry, had devoted himself to the study of robots with the combination of patience, audacity, lucidity and imagination of which his genius was composed.

He had first concentrated on improving the old calculating machines. As a result of his endeavours these had contrived to carry out simultaneously and with absolute accuracy a multitude of complicated operations involving thousands of figures and in a progressively shorter space of time. At this heroic period there had been fierce rivalry in this field and competitions were held every year at which the companies' various models came to grips. Thanks to Professor Fontaine's ingenuity, the EBC had always outclassed its rivals and its machines had established records which were expressed in millions of figures worked out in a fraction of a second.

In fact there appeared to be no theoretical limit to the performances accomplished by the professor's machines. The Board of Directors had had to confine him to a ceiling, above which the results were too abundant to be put to practical use. Once this ceiling was reached, the machines of the other companies had slowly attained it. But long before they did so, the scientist had immersed himself in fresh researches, no longer aimed at perfecting the old machines but at creating others, capable of solving problems of a more delicate nature than arithmetical operations.

He had obtained such results that the EBC's publicity had long since ceased to refer to 'calculation' and instead used the word 'thought'.

This claim had given rise to heated controversy. The objectors asserted that the most ingenious achievements in the field of cybernetics would always remain mechanical, that's to say they would never be able to solve anything but problems whose solution was explicitly or implicitly contained in the data. The machine, they said, combines the elements of these data and reproduces them in a different form known as result or conclusion. It was merely a 'formal' transformation and in no way a creative process comparable to that of the human mind.

But to this Professor Fontaine replied that properly speaking there was no such thing as 'creation', since this word should *always* be understood in the sense of 'combination' or potential rearrangement of former facts. According to him, in all operations of the human mind the solution or outcome was *always* contained, at least implicitly, in previous data. The brain had never done otherwise than modify the disposition of these data and present them in a new aspect. Consequently, between the aforesaid human brain and the artificial electronic brain there was a difference only of quality and not of nature. By a systematic improvement of the machine, he claimed, one would eventually be able to obtain all the manifestations generally regarded as being within the exclusive realm of intellect.

In support for this thesis he quoted several examples taken from extremely varied branches of human activity, frequently stretching his argument to the point of absurd paradox, and occasionally indulging in a childish play on words. He claimed, for instance, that there existed no difference in 'substance' between primary education and secondary, the latter being merely a modification in form of the former, just

as, in an electric transformer, the energy of the 'secondary' circuit is equal to that of the 'primary' in a new guise. Similarly, he saw in higher education a third presentation of the same fact and asserted that the three so-called degrees of mind corresponding to these three classes were differentiated only by their capacity to combine severally the immediate data of sensory knowledge.

He added:

'The number of cells in a human brain being finite, whereas the number of the circuits in a machine is unlimited, mere calculation of probabilities shows that machines will one day attain more complicated forms of arrangement than those within the scope of the human mind, and therefore more profound speculations and original "creations" (as you call them) exceeding human capacity.'

This postulate being acknowledged, he had set to work, both to verify his theories by practical experiment and to satisfy the demands of the Board of Directors of the EBC.

His first robot superior to the calculating machine was a scientific curiosity which gave him intense personal satisfaction but which enjoyed only mediocre success with the customers. It was a 'mathematician' and not a mere calculator. It was furnished, by means of a series of knobs and levers, with the fundamental postulates, the definition of the numbers and the symbols of analysis. Various electronic circuits combined these data and it then entered upon the multiple branches of this science, extending its deductions in each branch as far as the latest conclusions formulated by the greatest mathematical minds, and rediscovering the most famous theorems. By means of more and more complicated permutations and arrangements of the data, it had even reached the stage of stating theorems that had never before been formulated, the accuracy of which had been verified a *posteriori* but the demonstration of which had defeated the leading authorities.

Professor Fontaine's opponents had maintained that the so-called 'new' theorems had existed for all eternity in the basic axioms and definitions, in which point they were in agreement with the scientist. But they added that mathematical analysis, being precisely the science of formal transformations, involved only a minute and purely mechanical fraction of the intellect. The robot mathematician provided no proof as to the actual intelligence of a machine. The professor had shrugged his shoulders and pursued his research.

Since his employers wanted models capable of rousing widespread interest by means of spectacular demonstrations, he had concentrated his prodigious faculties on the game of chess. Several authorities on cybernetics had already been drawn to this subject, and some special situations, carefully prepared, which were played out with only three or four chessmen in a succession of well defined moves, had already been solved successfully by machines. This was obviously elementary. The genius of Professor Fontaine lay in tackling the problem as a whole.

That the so-called Maelzel automaton was a fake and that in actual fact a human being was concealed in the mechanical body, the Professor had readily admitted. The state of science at the time at which Maelzel was working led inevitably to this conclusion. But that the principle of the invention was Utopian, that a machine capable of playing chess was inconceivable, the scientist had fiercely denied. During the period of cogitation that preceded his undertaking, he had applied himself to destroying the arguments put forward against the possibility of such an achievement, reproving their authors for an unforgivable confusion between the qualifications 'infinite' and 'immensely large', accusing them of applying the former term implicitly and erroneously to the combinations of a game of chess.

Addressing his inner demon of contradiction, which permanently haunted him as it haunts all scientific minds, Professor Fontaine had said:

'Consider a game of chess at a given stage, which I call the 'initial moment'. It is White to play. The sum total of possible moves is high, I don't deny. But you must agree with me that it is finite and perfectly determined. To verify this, you need only take each white chessman one after the other and number the squares to which they can move according to the rules of the game. You must admit there's nothing simpler than to visualize a machine designed to carry out this purely mechanical operation.

'It is now Black to play. As *several* possible moves of the black chessmen correspond to *each* possible move of the white ones, it is obvious that the number of possible combinations after the second move is considerable. It is none the less obvious, however, that being the product of two perfectly determined factors, this number is likewise *finite*.

'Progressing by degrees, let us pass straight on, if you don't mind, to the nth move. You will see that the number of possible arrangements for the chessmen, albeit immensely large, and being expressed by a numerical symbol covering several pages, remains in every case absolutely definite. To say that the nth move depends entirely on the whim or skill of the player is a gross error. The latter has only the choice between a large number of determined moves and, after he has made a move, the consequent position is none other than a selection from an ineluctably fixed total.'

In his subsequent reasoning, the scientist enlarged ingeniously on the meaning of the word 'necessity.'

'So far,' he went on, 'I have considered all the combinations compatible with the displacement of the chessmen and the position of the squares. Now imagine a game between two beginners who know only the elementary moves. You will perceive that this total is already reduced by the tendency of both players to avoid a manifestly and immediately dangerous position such as would gratuitously place the queen in jeopardy or stupidly lay open the king. You call this the human factor. I say — and if you think it over, you'll readily agree — that the volition of the player in the act of eliminating these manifestly risky moves, which are characterized by a limited number of definite positions, may well be replaced, as to its effect, by a machine. A necessity of a mechanical order will take the place of the impulse induced by elementary prudence. The law will automatically avoid any combination resulting in the *immediate* loss of the game.

'Now observe two expert players. The number of favourable combinations is limited in a far higher proportion by what you call their skill and which is again related, I maintain, to a sort of mechanical necessity, even though this characteristic is less distinct than in the first case — the necessity of avoiding disaster in the future which does not appear clearly to the beginner because his brain is not trained to consider a large number of positions. The necessity will reveal itself to him only after the fifth move, for instance, but it is nevertheless contained in its entirety, in a potential state, in the fatal moves. If the opponents are first-class players, the limitation of their judgement will be even more pronounced (the necessity more imperative), being based on the more distinct, keener perception of arrangements possible in a more distant future. If you consult a champion on this matter, he will tell you that for him there exists only a small number of advantageous moves — what I call the only moves *possible* for him as a champion.

'I maintain,' Professor Fontaine had concluded, after pondering the question at great length, 'I maintain that a superchampion, or the robot I intend to build, must in every case reduce this number of possible moves to a single one, by automatically considering all the subsequent arrangements resulting from a certain position, right up to the end of the game, and by eliminating every fatal move. I intend to construct a robot which, being furnished with data corresponding to some position or other in a game of chess, will resolve the following problem — how to determine and make *the* right move, that's to say the one which contains no potential danger of losing.'

The application of this theory had led to astonishing results. The professor had created a robot which did indeed always make the right move and invariably beat any opponent. But this success had galvanized the research workers of the rival companies. One of them, working on similar principles, had then produced a second robot as ingenious as the first which likewise made no mistakes and invariably countered *a* good move by *the* good move. It then became apparent that all the possible games of chess were reduced to a single one: the ideal game, always identical with itself, and which invariably resulted, by an identical process, in a draw. Chess had thus been deprived of a great deal of its interest and the craze for electronic players had come to an end. Professor Fontaine had re-immersed himself in his speculations to discover something new.

'Mechanical,' Professor Fontaine's opponents had said, referring to the chess-player. 'Mechanical indeed,' the scientist had readily acknowledged, 'but everything human is mechanical — language, for instance.' And without heeding the sarcastic remarks that had ridiculed these claims or the songs that celebrated the coming of a robot-writer, he had gone more deeply into this idea by thrashing it out with his own demon.

'For the time being,' he had said, 'there is no question of composing a story, even though this should not be dismissed as a future possibility. . . . We shall make a modest start and confine ourselves to a few short sentences.

'It is not particularly difficult to imagine a machine capable of producing words. Words are only combinations, finite in number, of consonants and vowels, and their synthesis may be reduced to a purely mechanical operation starting from an initial letter. Consider a robot which is furnished with data consisting of a very large number of these consonants and vowels. It is child's play to imagine an apparatus connecting every word in the language to the machine and in such a way that all non-existing combinations of letters are automatically eliminated. Supposing that the first circuit happens to start off with the letter *s*. A second circuit then comes into action and proposes *d* as the second letter. The combination *sd* is impossible as the first two letters of a word. It will be rejected by the apparatus, the *d* will be discarded, and the machine will successively suggest other letters until one of these is admissible. If *h* is suggested it will be admitted. The combination *sh* will be selected; further circuits will yield a third letter, and so on. Starting from the fortuitous initial letter, then proceeding by trial and error and without intervention from the human mind, the robot will compose a group necessarily corresponding to an existing word beginning with *sh* — "sheep" for instance. When such a combination has thus been "found", then, and only then, will my apparatus automatically halt the process.

'It is equally easy to imagine specialized compartments for the formation of nouns, adjectives, verbs, articles and other parts of speech, singulars and plurals, persons and tenses. We may then conceive a second stage at which an initial conjunction of words will be made. If the noun "sheep" has been chosen, it is obviously within the possibilities of a mechanism to combine it with a grammatically suitable article and adjective — to select, for instance, phrases such as "the fluid sheep", or "the sombre sheep" or "the white sheep", to the exclusion of any that conflict with the strict rules of agreement in gender and number. Up to this point we have encountered no difficulty. . . .'

' "The fluid sheep" doesn't make sense,' interjected the demon of contradiction.

'Let me go on. Everything in its own time. We shall not find much more difficulty at the following stage, which will see the complete construction of a simple sentence in agreement with the rules of syntax, which are determined and which a machine can apply just as well as and even better than a human brain. We shall thus see the formation of a certain number of grammatically correct phrases such as "the fluid sheep flies under the pointed sky", or "the white sheep eats grass". . . .'

'That's what I'm getting at,' the demon once more broke in. 'Most of your sentences, though grammatically correct as you say, will be absurd.'

'They will be irreproachable from the point of view of construction. That's the essential point, which I wanted to make you admit. Among them there are bound to be some that will make sense. We shall then only need to proceed with a fresh selection. This is where my theory of "primary truths" comes in.'

'Primary truths?'

'Follow closely. The problem is of the same order as that of the game of chess, albeit slightly more subtle.

'In this case it's a question of eliminating any expression that is meaningless. You think perhaps this is beyond the capacities of a machine? Nothing of the sort. Every meaning is merely a transformation of a previous meaning. The most erudite pronouncement is merely a permutation of the elementary findings of a primary pronouncement. Proceeding by degrees, we arrive at a statement of fact, or what I call a "primary truth". These are limited in number and may well be electronically connected to the selective organs of a machine which, after carrying out the necessary permutation, will ruthlessly eliminate from the expressions it is given anything that conflicts with any of these statements or even anything that is not connected with any of them. Only the sentences which really make sense will be produced. I maintain that the human mind does not proceed in any other way when it tries to express itself.'

The professor had built his machine. The construction of his sense selector had seemed a tricky business at first. But in the process of determining the primary truths, he had noticed that their number was less great than he had supposed. It needed only a suitable classification to reduce them all to a few fundamental axioms such as 'A is A' or 'A is not non-A', and to a small number of immediate sense perceptions. Starting from these data alone, and by combining them, the machine managed to discern any expression that did not make sense. Thus sentences such as 'the fluid sheep flies under the pointed sky' or 'the white sheep eats veal' were eliminated. Partial groups such as 'fluid sheep' or 'pointed sky' were discarded by a preliminary selector. A verb like 'fly' or an outlandish complement like 'veal' were rejected by the final apparatus, until finally there appeared a combination such as 'the white sheep walks under the blue sky' or 'a white sheep eats grass', statements which were compatible with all the primary truths and which the robot produced automatically.

The ingenious professor had devoted several years to perfecting his electronic writer.

' "Progress" in this field,' he explained, 'consists in obtaining expressions which, even though being transformations of primary truths, as they all are, carefully conceal this fact under the complexity or the originality of their construction, thereby demanding a lengthy process of unravelling to arrive at the source.'

Thus it was that, having started from statements of fact such as 'the white sheep is white', then, after being given the translation of the various data in several languages, having continued with 'Ego sum qui sum', the robot writer had gradually elaborated its expressions and managed to produce more complex statements. It had produced: 'Ex nihilo nihil fit', 'The cat eats the mouse', 'Nevertheless it turns', 'I think, therefore I am', and one day, which was a red-letter day for Professor Fontaine, 'To be or not to be, that is the question'.

The scientist's rivals having likewise contrived to build robot writers, and the EBC still wishing to maintain its lead in this field, he had attempted to reproduce other features of human activity by mechanical means. Since circulation, respiration, alimentation and digestion had long since been achieved by machines, he had aimed higher and made an intensive study of the laws of sexual attraction and of the biological factors, considered as initial data, which had as their outcome the carnal act. He had concluded that there was nothing to prevent robots from having sexual intercourse.

He had given an experimental demonstration of this. Several male and female automata, which he had constructed, had been placed at random in a room. At first they had wandered about aimlessly, then, after a certain length of time, appeared to be attracted one to another as though by a magnetic force. The attraction always occurred between robots of the opposite sex, and the couples went through the movements of sexual intercourse with a realism which caused intense emotion among the directors of the EBC when they first witnessed the spectacle. For these automata, to which he had given a human aspect, not only imitated to perfection all the attitudes of living creatures but also, as he had foreseen, invented new combinations thanks to the subtle complexity of the electronic brain which controlled their movements.

As Professor Fontaine foresaw, these latest achievements were to lead in the long run to 'sentiment'. But above all they gave him the idea for a new machine which would mark one of the most important stages of cybernetics, by reducing to nought the objections of its everlasting detractors.

'Your mechanisms,' the latter said in so many words, 'are the most ingenious in the world, we don't deny it. But they merely testify to your scientific knowledge, to your inventive genius. All the qualifications of these robots must be attributed to you, who created them. As we have always maintained, they prove the power of human intellect and nothing else.'

The professor had then had the brainwave of constructing a robot capable of engendering other robots. He had succeeded, as in everything else he had so far attempted. Combining this new invention with the previous one, he had constructed a machine in the depths of which, after sexual intercourse, there developed a cell which grew and multiplied by a mechanical process, became the foetus of a robot, was born and grew up, eventually appearing as a machine similar to its parents.

When he had verified this marvel, the professor had intoned a hymn of victory, but his opponents had not declared themselves beaten. They had said:

'It's all very well, but the creator of this uterus is still you — you, a human being.'

Then the scientist had constructed a machine capable itself of engendering a uterus and, to save time and forestall a lot of pointless argument, he had straight away created a robot capable of perpetuating itself indefinitely through an unlimited series of descendants. After which, at a conference which had made history, he had demonstrated beyond all shadow of doubt that any term in this series was strictly equivalent to zero in relation to the infinite succession of future creatures, that, since his own role could not be eliminated, the creative spirit should consequently be attributed to the robot considered as an eternal entity.

His detractors had been unable to think of any reasonable objection to this demonstration. They had not admitted defeat, however, and had taken refuge behind the inexpressible. They had shaken their heads and stubbornly declared:

'Call them anything you like, but your robots are not human. They lack some element or other which we are unable to define but of which we sense the absence.'

At this, Professor Fontaine had been deeply disturbed. His own demon of contradiction, never so powerful as at this time, kept repeating over and over again:

'They're right. Your robots are not human. They lack some indefinable characteristic.'

He had feverishly set to work again and, with a final effort, had succeeded in creating a sort of mechanism, the successive generations of which 'evolved'. This evolution, which followed a strict course, was slow but nevertheless perceptible thanks to the extreme speed of reproduction which the scientist had bestowed on the species. In this way he artificially obtained robot birds from robot reptiles, robot mammals from robot fish, and robot men from robot primates, without any intervention on his part other than the initial regulation of the mechanism. He went further. Starting from robot men, and by combining the organs in two different ways, he produced two species, which evolved in a clear-cut manner, on the one hand towards Good, on the other towards Evil.

After performing these miracles, the professor once again questioned his inner demon.

'You're not there yet,' the latter replied.

It then looked as though the genius of Professor Fontaine had reached its limits. Several years went by, during which he exhausted himself in fruitless endeavours to take a step forward and eliminate the subtle difference which still separated, as he was now convinced, his machines from human beings. The Board of Directors of the EBC were worried at not seeing any fresh marvels emerge from his laboratory which was once so fecund, and some of them began to mutter that he was somewhat too old to hold such an important position.

This year, at last, breaking a long silence, the scientist had declared he had an announcement of the greatest importance to make. The Board had met.

'*Eureka,*' the professor said again.

'Go on,' said the Board with bated breath.

'I have found,' the scientist translated.

'Yes, but what?' the Board enquired.

'The ideal machine. The human machine. The solution was there all the time, staring me in the face, within the reach of the veriest ignoramus. It had escaped me!'

'Go on, go on!' said the Board.

'Listen, gentlemen,' said the man of science in a voice trembling with emotion. 'No one can have any further objection to raise against my robots. I now possess the secret which will crown my career of arduous labour and meditation. I can give you palpable proof of my theories. My machine now acts exactly like a human being. It "thinks" like a human being. It possesses an intellect similar to our own. This characteristic was already contained at the latent stage in all my previous inventions. It needed only a minute detail, which was too simple — that's why it had escaped me. To endow my robots with this indefinable quality which they still lacked, gentlemen, I merely had to . . .'

'What?' yelped the Board.

'*Sabotage* them, gentlemen,' the prince of science bellowed in triumph.

'Sabotage them!' exclaimed the Board, dreadfully disillusioned and convinced that their great man had taken leave of his senses.

'Sabotage them,' Professor Fontaine insisted vehemently. 'That's what I've done. I have *unhinged* them. Do you understand? My calculators, for instance, now make mistakes. They no longer invariably give the accurate answer. Do you follow? They go wrong. They only occasionally produce a correct solution.'

This announcement was greeted by a long contemptuous silence. Then the Board thought it over and asked to see the prototypes. Professor Fontaine produced them. He first of all showed his calculating machines which gave wrong results. At this sight the more sensitive directors experienced an incipient pang of emotion.

But the enthusiasm grew general and turned to delirium when the scientist exhibited his other models, which he had built according to the same principle. So much so that when he had revealed the prodigiously 'human' characteristics with which they had been endowed, the Board, ashamed of ever having doubted him, apologized profusely to Professor Fontaine and unanimously voted the necessary funds for the mass production of the perfect robots.

The mathematician went astray in a maze of contradictions and only exceptionally arrived at an answer. The chess player lost most of the games it played. The robots in love confused the sexes. The evolutive species oscillated between incoherent states, so that it was impossible to tell if they tended towards Good or Evil.

As for the electronic writer, Professor Fontaine had merely had to do away with its primary truth mechanism, its sense selector, for it to produce to its heart's content such sentences as 'The fluid sheep flies under the pointed sky' or 'the white sheep eats veal', manifestations to which the fiercest detractors were obliged to yield, acknowledging at last the appearance of the final human characteristics that were still lacking: an artistic sense and a sense of humour.

"*Sometimes I ask myself, 'Where will it ever end?'*"

(*Drawing by Chas. Addams;* © 1946, 1974 *The New Yorker Magazine, Inc.*)

BIBLIOGRAPHY

MOVIES AND MACHINE ART

Bibliography

ROBOTS IN FICTION, POETRY, AND DRAMA: THE BASIC LIST

Aldiss, Brian W. *Who Can Replace a Man?* New York: New American Library, 1967.

Andersen, Hans. "The Emperor's Nightingale," in *The Complete Andersen.* New York: The Limited Editions Club, 1949.

Asimov, Isaac. *The Caves of Steel.* New York: New American Library, 1955.

————. *I, Robot.* Greenwich, Conn.: Fawcett Crest Books, 1970. (Nine short stories.)

————. *The Naked Sun.* Garden City, N.Y.: Doubleday, 1957.

————. *The Rest of the Robots.* Garden City, N.Y.: Doubleday, 1964. (Eight short stories plus the two novels: *The Caves of Steel* and *The Naked Sun.*)

B., Madam. *La Femme Endormie.* Paris, 1899.

Bates, Harry. "Farewell to the Master," in Raymond J. Healy and J. Francis McComas, eds., *Famous Science-Fiction Stories.* New York: Random House, 1946. (Source of the film *The Day the Earth Stood Still.*)

Bierce, Ambrose. "Moxon's Master," in *The Complete Short Stories of Ambrose Bierce.* Garden City, N.Y.: Doubleday, 1970.

Biggle, Lloyd, Jr. *The Metallic Muse.* New York: Daw Books, Inc., 1972.

Binder, Eando. "Adam Link's Vengeance," in Phil Stong, ed., *The Other Worlds.* New York: Wilfred Funk, 1941.

————. *The Double Man.* New York: Modern Literary Editions Publishing Co., 1971.

————. "I, Robot," in *Amazing,* January 1939.

Blish, James, and Robert Lawndes. *The Duplicated Man.* New York: Airmont Publishing Co., 1964.

Bloch, Chaim. *The Golem: Legends of the Ghetto of Prague,* translated from the German by Harry Schneidermann with a prefatory note by Hans Ludwig Held. Vienna: The Golem, [1925].

————. *The Golem: Mystical Tales from the Ghetto of Prague.* New York: Rudolf Steiner Publications, 1972.

————. *Israel der Gotteskampfer; der Baalschem von Chelm und sein Golem, ein ostjudisches Legendbuc.* Berlin: B. Harz, 1920.

————. *Der Prager Golem, von swiner 'Geburt' bis zu seinem 'Tod.'* Vienna: Dr. Blochs Woschenschrift, 1919.

Boulle, Pierre. *Time Out of Mind.* New York: New American Library, 1969. (Twelve short stories, including "The Perfect Robot" and "The Man Who Hated Machines.")

Bounds, Sydney J. *The Robot Brains.* New York: Macfadden-Bartell, 1969.

Bradshaw, William Richard. *The Goddess of Atvatabar.* New York: J. F. Douthitt, 1892.

Burroughs, Edgar Rice. *The Synthetic Men of Mars.* Tarzana, Calif.: Burroughs, 1940.

Butler, Samuel. *Erewhon.* London: Trübner, 1872.

Butti, A. E. *L'Automate.* Paris: Société du Mercure de France, 1898.

Caidin, Martin. *Cyborg.* Warner Paperback Library, 1972. (Source of the TV series, *The Six Million Dollar Man.*)

Čapek, Karel. *R.U.R.* (*Rossum's Universal Robots*), *a Fantastic Melodrama,* translated by Paul Selver. The Theatre Guild version with four illustrations from photographs of the Theatre Guild production. Garden City: Doubleday, 1923.

Conklin, Groff, ed. *Science Fiction Thinking Machines.* New York: Vanguard Press, 1954. (Short stories by S. Fowler Wright, Ambrose Bierce, Isaac Asimov, Raymond A. Gallun, Fritz Leiber, Harbert Goldstone, Eric Frank Russell, William Tenn, Clifford D. Simak, Michael Shaara, Theodore Sturgeon, etc. Also Čapek's play, *R.U.R.*)

Correia, Natalia. *O Homunculo: Tragedia.* [Lisbon:] Contraponto, [1964]. (Play.)

Crichton, Michael. *Westworld.* New York: Bantam Books, 1974. (Screenplay of the film.)

Davidson, Avram. "The Golem," in Judith Merril, ed., *Year's Greatest Science Fiction and Fantasy: vol I.* New York: Gnome, 1956.

Delblanc, Sven. *Homunculus. En magisk berättlst.* Stockholm: Bonnier, 1965.

del Rey, Lester. "Helen O'Loy." *Astounding,* December 1938.

Dick, Philip K. *Do Androids Dream of Electric Sheep?* Garden City, N.Y.: Doubleday, 1968.

————. *We Can Build You.* New York: Daw Books, Inc., 1972.

Dunsany, Lord. *The Last Revolution.* London & New York: Jarrolds, 1951.

Elwood, Roger, ed. *Invasion of the Robots.* New York: Paperback Library, 1969. (Short stories by Isaac Asimov, Robert Bloch, Philip K. Dick, Lester Del Rey, Richard Matheson, Eric Frank Russell, and Jack Williamson.)

Ewers, H. H. *Alraune.* Berlin Zehlendorf: Sieben Stäbe-Verlag, 1928.

Fairman, Paul W. *I, the Machine.* New York: Lodestone Books, 1968.

Feiffer, Jules. "The Lonely Machine," in *Feiffer's Album.* New York: Random House, 1963.

Ferman, Edward L., and Barry N. Malzberg, eds., *Final Stage: The Ultimate Science Fiction Anthology.* New York: Penguin Books, 1975. (Includes a robot story by Asimov and a story by Malzberg about an uncontrollable machine.)

Ferry, Jean. *Le Mécanicien et autres contes.* Paris: Gallimard, 1953.

Gavin, Thomas. *Kingkill.* New York: Random House, 1977. (A novel about Maelzel and his chess-playing dwarf.)

Gerrold, David. *When Harlie Was One.* New York: Ballantine Books, 1972.

Gmeyer, Anna. *Automatenbuffett: ein Spiel in drei Akten mit Vorspiel und Nachspiel.* Berlin: Arcadia Verlag, [1932?]. (Play.)

Goulart, Ron. *Broke Down Engine and Other Troubles with Machines.* New York: Collier-Macmillan, 1971.

Greenberg, Martin. *The Robot and the Man.* New York: Gnome, 1953. (Short stories by John D. MacDonald, Bernard Wolfe, Lewis Padgett, J. J. Holmes, John Browning, A. E. van Vogt, Lester del Rey, Joseph E. Kelleam, and Robert Moore Williams.)

Hamilton, Edmond. "The Metal Giants." *Weird Tales,* December 1926.

Harbou, Thea von. *Metropolis.* London: The Readers Library, 1927; New York: Ace Books, 1927. (Novel on which Fritz Lang's film is based.)

Harrison, Harry. *War with the Robots.* New York: Pyramid, 1962.

Hess, Johannes. *Der Rabbiner von Prag* [Rabbi Loew]: *Kabbalistisches Drama in vier Akten nach einer Prager Legende.* Karlsruhe: Fr. Gutsch, [1914]. (Play.)

Hoffmann, E. T. A. "Automata," "The Sandman," in *The Best Tales of Hoffmann,* ed. E. F. Bleiler. New York: Dover Books, 1967.

Huxley, Aldous. *Brave New World.* Garden City, New York: Doubleday, Doran & Co., 1932.

Jones, Raymond F. *The Cybernetic Brains.* New York: Paperback Library, 1969.

Kastle, Herbert D. *The Reassembled Man.* Greenwich, Conn.: Fawcett Publications, 1964.

Klein, Karl. *Wider den Golem: ein Beitrag zur Frage der Gegenwärtigkeit des Menschseins.* Dortmund: Borgmann, 1960.

Knight, Damon, ed. *Science Fiction Inventions.* New York: Lancer Books, 1967. (Short stories by Cordwainer Smith, Harry Harrison, Hanry Kuttner, Katherine MacLean, Theodore Sturgeon, L. Sprague de Camp, Isaac Asimov, John Pierce, Carol Emschwiller, and Frank Herbert.)

Kuttner, Henry, *Return to Otherness.* New York: Ballantine Books, 1962.

————. *Robots Have No Tails.* New York: Gnome, 1952.

Lem, Stanislaw. *Mortal Engines.* New York: Seabury Press, 1977.

Lewis, Arthur O., Jr. *Of Men and Machines.* New York: E. P. Dutton & Co., 1963. (Important anthology of drama, fiction, poetry, and nonfiction dealing with the machine as enemy, friend, and thing of beauty. Special sections are devoted to material on problems of the machine and future technological developments. The book includes the complete text of Čapek's play, *R.U.R.*)

Long, Frank Belknap. *It Was the Day of the Robot*. New York: Belmont Books, 1963.

McMullen, Joseph Carl. *The Automatic Butler: a farce in Three Acts*. Boston: Walter H. Boker Company, [1929].

Merritt, A. *The Metal Monster*. New York: Avon Books, 1946.

Meyrinck, Gustav. *The Golem*. Prague and San Francisco: Mudra, 1972.

Moskowitz, Sam, ed. *The Coming of the Robots*. New York: Collier Books, 1963. (Short stories by Eano Binder, Lester del Rey, John Wyndham, Isaac Asimov, Clifford D. Simak, etc.)

Mowshowitz, Abbe, ed. *Inside Information: Computers in Fiction*. Reading, Mass.: Addison-Wesley, 1977. (Important anthology with detailed bibliography.)

Nolan, William F., ed. *The Pseudo-People: Androids in Science Fiction*. New York: Berkley Publishing Corporation, 1965. (Short stories by Henry Kuttner, Ray Bradbury, Shelly Lowenkopf, Chad Oliver, Isaac Asimov, James Causey, Charles E. Fritch, Dennis Etchison, Richard Matheson, Ron Goulart, Charles Beaumont, Frank Anmar, and William F. Nolan. The book contains a bibliography of robot fiction — listing mainly short stories.)

Olemy, P. T. *The Clones*. New York: Caravelle Books, Inc., 1968.

Rambo, Ana M. Feroleto. *Los Robots; fragmento*. [Buenos Aires]: Editorial Cuadernos del siroco, 1962. (Poetry.)

Raymond, Alex. *Flash Gordon: The War of the Cybernauts*. New York: Avon Books, 1975.

Richter, Jean-Paul. *The Death of an Angel and Other Pieces*. London: Black and Armstrong, 1839.

Shaw, Bernard. *Back to Methuselah*. London: Constable, 1921. (Part V of Shaw's longest play contains a memorable episode involving two synthetic figures and their inventor.)

Shelley, Mary W. *Frankenstein; or The Modern Prometheus*, edited with an introduction by M. K. Joseph. London, New York: Oxford University Press, 1971.

Silverberg, Robert, ed. *Men and Machines*. New York: Award Books, 1968. (Short stories by Brian W. Aldiss, James Blish, Lester del Rey, Randall Garrett, Lewis Padgett, Fred Saberhagen, Robert Silverberg, George O. Smith, and Jack Williamson.)

Simak, Clifford D. *City*. New York: Gnome, 1952.

———. *Time and Again*. New York: Simon & Schuster, 1951.

Stapledon, W. Olaf. *Last and First Men*. London: Methuen & Co., 1930.

Sturgeon, Theodore. *More than Human*. New York: Farrar, Straus & Young, 1953.

Veuzit, Max du [pseud. for Alphonsine Simonet]. *L'Automate*. Paris: Espes, [1935?].

Villers de l'Isle Adam, Jean Marie Matthias . . . Comte de. *L'Eve Future*. Paris: M. de Brunhoff, 1886.

Vonnegut, Kurt, Jr., *Player Piano*. New York: Scribner's, [1952].

Williams, Robert Moore. "Robots' Return." *Astounding*, September 1938.

Williamson, Jack. *The Humanoids*. New York: Lancer Books, 1963.

———. *The Pandora Effect*. New York: Ace Books, 1969. (Seven short stories, including "The Metal Man" and "With Folded Hands . . .".)

———. *People Machines*. New York: Ace Books, 1971.

Wolfe, Bernard. *Limbo*. New York: Random House, 1952. British title: *Limbo 90*. London: Secker & Warburg, 1953.

Zollinger, Jacob. *Homunculus: ein Drama in funf Akten*. Aaru: J. R. Sauerlander & Co., 1924. (Play.)

Robots in Nonfiction and Technical Studies

Annan, David. *Robot The Mechanical Monster*. New York: Bounty Books, 1976.

Anon. "The Jaquet-Droz Androids." *Scientific American* 88 (April 18, 1903): 301–302.

Anon. *The Famous Chess-Player* [of W. R. von Kempelen]; no. 14, St. James's Street, next Brooks's. [London]: Printed by H. Reynall, [1783?].

Anon. *Il segreto del famoso automa che giucava a scacchi*. Florence: G. Benelli, 1841.

Arbib, Michael. "Man-Machine Symbiosis and the Evolution of Human Freedom." *The American Scholar* 43 (Winter 1972–1973).

Archer, Mildred. *Tippoo's Tiger*. London: H. M. Stationery Office, 1959.

Arrington, Joseph Earl. "John Maelzel, Master Showman of Automata and Panoramas," *Pennsylvania Magazine of History and Biography*, vol. 84, no. 1 (January 1960), 56–92.

Basserman-Jordan, Ernst von. *Die Wasser-Automaten und Wasserkunste im Parke des Lustschlosses Hellbrunn bei Salzburg*. Leipzig: W. Diebener, G.m.b.H., [1928].

Berkeley, Edmund Callis. *The Construction of Living Robots*. New York: E. C. Berkeley, 1952.

Berkeley, Edmund Callis, and Jack Koff. *Squee, the Robot Squirrel — Construction Plans*. New York: E. C. Berkeley, 1952.

Boehn, Max von. *Puppets and Automata*, translated by Josephine Nicoll, with a note on puppets by George Bernard Shaw. New York: Dover Publications, 1972.

Bruinsma, A. *Multivibrator Circuits; Introduction to Robot Technique*, translated from the German by E. Harker. New York: Macmillan, 1960.

———. *Practical Robot Circuits: Electronic Sensory Organs and Nerve Systems*, translated from the German by E. Harker. Eindhoven: Philips, 1959.

Burks, Arthur Walter, ed. *Essays on Cellular Automata*. Urbana: University of Illinois Press, 1970.

Calder, Ritchie. *The Evolution of the Machine*. Wisconsin: E. M. Hale and Company for the Smithsonian Institution, 1968.

Carroll, Charles Michael. *The Great Chess Automation*. New York: Dover Publications, 1975.

Chapuis, Alfred. *Automates: Machines automatiques et Machinisme*. Pref. de Alphonse Bernoud. Geneva: S. A. des Publications Techniques, 1928.

———. *Les Automates dans les Oeuvres d' Imagination*. Neuchâtel: Editions du Griffon, [1947].

———. *Les Automates, figures artificielles d'hommes et d'animaux; histoire et technique*. Neuchâtel: Editions du Griffon, [1949].

———. "Du Goût des Chinois pour les Automates," *Journal Suisse d'Horlogerie et de Bijouterie* 46 (1921): 154–155.

———. *The Jacquet-Droz Automatons*. Neuchâtel: History Museum, 1956.

Chapuis, Alfred, and Edmond Droz. *Automata: A Historical and Technological Study*, translated by Alec Reid. Neuchâtel: Editions du Griffon, 1958.

Chapuis, Alfred, and E. Gelis. *Le Monde des Automates: étude historique et technique*. 2 vols. Paris: [Blondel la Rougery], 1928.

Cohen, John. *Human Robots in Myth and Science*. London: G. Allen & Unwin, [1966].

Culbertson, James Thomas. *The Minds of Robots: Sense Data, Memory Images, and Behavior in Conscious Automata*. Urbana: University of Illinois Press, 1963.

Devaux, Pierre. *Automates et Automatisme*. Paris: Presses Universitaires de France, 1944.

Diels, Herman. *Antike Technik: Sieben Vorträge*. Leipzig: B. G. Teubner, 1920.

Droz, E. "From Jointed Doll to Talking Robot." *New Scientist*, vol. 14, no. 282 (1962), pp. 37–40.

Evans, Henry Ridgely. *Edgar Allan Poe and Baron von Kempelen's Automaton*. Kenton, Ohio: International Brotherhood of Magicians, 1939.

Florescu, Radu. *In Search of Frankenstein*. New York: Warner Books, 1976.

Freudenthal, Hans. *Machines Pensantes. Conférence faite au Palais de la Découverte le Mai 1953*. [Paris, 1953].

Gerber, A. "Goethe's Homunculus," *Modern Language Notes* 12 (1897): 69–79.

Giedion, Siegfried. *Mechanization Takes Command*. New York: W. W. Norton, 1969.

Gill, Arthur. *Introduction to the Theory of Finite-State Machinery*. New York: McGraw-Hill, 1952.

Ginzburg, Abraham. *Algebraic Theory of Automata*, New York: Academic Press, 1968.

Ginsburg, Seymour. *An Introduction to Mathematical Machine Theory*. Reading, Mass.: Addison-Wesley Publishing Co., [1962].

Glut, Donald F. *The Frankenstein Legend*. Metuchen, N.J.: The Scarecrow Press, 1973.

Guilbaud, G. T. *What is Cybernetics?*, translated by V. Mackay. London: Heinemann, 1961.

Haas, Walter de. *Automaten: die Befreiung des Menschen durch die Maschine*. Stuttgart:: Dieck & Co., [1930].

Hatfield, Henry Stafford. *Automaten: or The Future of the Mechanical Man.* London: K. Paul, Trench, Trubner & Co., Ltd., 1928.

Held, H. L. *Das Gespenst des Golem.* Munich: Allgemeine Verlagsanstalt, 1927.

Hero of Alexandria, *Di Herone Alessandrino De gli automati . . . tradotte dal greco da Bernardino Baldi.* Venice: Appresso Girolamo Porro, 1589.

Hu, Sze-Tsen. *Mathematical Switching Circuits and Automata.* Berkeley: University of California Press, 1968.

J.A. "Automatas, cajas de musica y pajaros mecanicos." *Revista de las artes y los oficios,* no. 75 (1950), pp. 24–28.

Keller, David Henry. *The Homunculus.* Philadelphia: Prime Press, 1949.

Kiaulehn, Walther. "Automaten-Raritaten." *Der Türmer,* Berlin, February 1936, pp. 449–452.

Knips-Hasse, A. *Das Automaten-Kabinett.* Berlin: E. Bloch, [1893].

Krijgelmans, C. C. *Homunculi.* Amsterdam: De Bezige Bij, 1967.

Le Bot, Marc, et al. *Le Macchine Celibi/The Bachelor Machines.* New York: Rizzoli, 1975.

Lewis, Arthur O., Jr., ed. *Of Men and Machines.* New York: E. P. Dutton, 1963.

Ludwig, Albert. "Homunculi und Androiden." *Archiv fur das Studium der neueren Sprachen und Literaturen.* Braunschweig, 1918. Band 137, pp. 137–153; Band 138, pp. 141–155; Band 139, pp. 1–25.

Maingot, Elaine. *Les Automates.* [Paris]: Hachette, [1959].

Martin, Henri. "Historique des Automates du Moyen Age à la Fin du XVIIme siècle." *Journal Suisse d'Horlogerie.* 32 (1908): 244–248, 316–321, 382–389, 419–422; 33 (1909): 60–65, 98–103.

Maskelyne, John N., and George A. Cooke. *A Guide to their . . . entertainment of modern miracles.* London, [187?].

Moore, Edward F., ed. *Sequential Machines.* Reading, Mass.: Addison-Wesley Publishing Co., [1964].

Nelson, Raymond John. *Introduction to Automata.* New York: Wiley [1967].

Neumann, John von, and Arthur W. Burks.

Theory of self-reproducing Automata. Urbana: University of Illinois Press, 1966.

Papp, Desiderius. *Der Maschinenmensh.* Wien: Stein, 1925.

Perregaux, Charles. *Les Jaquet-Droz et leurs Automates.* Neuchâtel: Wolfrath & Sperle, 1906.

Petzoldt, Fritz. *Werkzeugeinrichtungen auf Einspindelautomaten,* Berlin: J. Springer, 1941.

Pontus-Hultén, K. G. *The Machine as seen at the End of the Mechanical Age.* Greenwich, Conn.: New York Graphic Society, 1968.

Prasteau, Jean. *Les Automates.* Paris: Grund, [1968].

Racknitz, Joseph F. *Ueber den Schachspieler des Herrn von Kempelen und dessen Nachbildung.* Leipzig and Dresden: Breitkopf, 1789.

Rosenfeld, B. *Die Golemsage und ihre Verwertung in der deutschen Literatur.* Breslau: Priebatsch, 1934.

Rosenfield, L. C. *From Beast-Machine to Man-Machine.* New York: Oxford University Press, 1940.

Sablière, Jean, ed. *De L'Automate à L'Automatisation. Textes de Héron d'Alesandre, Salomon de Caus, Camus La Lorrain, Jacques de Vaucanson,* [etc.]. Paris: Gauthier-Villars, 1966.

Sadoul, Jacques. *Hier, l'An 2000.* Paris: Denoël, 1973.

Salomaa, Arto. *On Probabilistic Automata with One Input Letter.* [Finland]: Turku, 1965.

Schuh, J. F. *Principles of Automatation: what a Robot can and cannot do.* Princeton, N.J.: Van Nostrand, 1965.

Scriven, Michael. "The Compleat Robot: A Prolegomena to Androidology," in Sidney Hook, ed., *Dimensions of Mind: A Symposium.* New York: New York University Press, 1960, pp. 118–142.

Shannon, Claude E., and J. McCarthy, eds. *Automata Studies.* Princeton, N.J.: Princeton University Press, 1956.

Silberer, Herbert. "Der Homunculus." *Imago* 3 (1914): 37–79.

Steiner, Hugo. "Der Golem. Prager Phantasien." *Jahrbuch Deutscher Bibliophilen und Literaturfreunde* 5 (1917): 114.

Strehl, Rolf. *The Robots are Among Us,* translated from the German by Herman Scott. London, New York: Arco Publishers, [1955].

239

Swoboda, Helmut. *Der Künstliche Mensh.* [Munchen]: Heimeran, 1967.

Torres y Quevedo, Leonardo. "Essais sur l'automatique." *Revue generale des Sciences Pures et Appliques* 26 (1915): 601–611.

Vaucanson, Jacques de. *Le Mécanisme du Fluteur Automate, présenté à Messieurs de l'Academie Royale des Sciences . . . Avec la description d'un canard artificiel, mangéant, buvant, digérant et se vuidant, épluchant ses ailes et ses plumes, imitant en diverses manières un canard vivant . . . Et aussi celle d'une autre figure . . . jouant du tambourin et de la flute, suivant la relation mecanisme du fluteur automate . . .* Paris: Chez Jacques Guerin, 1783.

Valtenin, Veit. "Goethes Homunkulus." *Modern Language Notes* 13 (1898): 432–443, 462–471.

Wiener, Norbert. *The Human Use of Human Beings: Cybernetics and Society.* Garden City, N.Y.: Doubleday, 1954.

Willis, Robert. *An Attempt to Analyze the Automaton Chess Player of Mr. de Kempelen.* London: J. Booth, 1821.

Windisch, Karl Gottlieb von. *Lettres sur le joueur d'echecs de M. Kempelen.* Basle: Chez l'Editeur, 1783.

Winograde, S., and J. D. Cowan. *Reliable Computation in the Presence of Noise.* Cambridge, Mass.: M.I.T. Press, 1963.

Robots in the Cinema

ANDROIDS, AUTOMATA AND ANIMATED PUPPETS
FROM THE BEGINNINGS OF FILM TO THE PRESENT

1897   *The Clown and the Automaton* (Fr.), Méliès film

1899   *L'Illusioniste Fin de Siècle* (Fr.), Méliès film (on the Coppélia theme)

1900   *Coppélia* (Fr.), Méliès film

1906   *The Motor Valet* (Br.), dir. Arthur Cooper

1907   *The Doll's Revenge*, dir. unknown

1908   *An Animated Doll*, dir. unknown

1909   *The Electric Servant* (Br.), dir. Walter Booth
Noah's Ark (Br.), dir. Arthur Cooper(?) (toys come to life in a child's dream)
Rubber Man, dir. Sigmund Lubin(?)
The Automatic Monkey (Fr.), dir. unknown

1910   *Automaton* (Fr.), dir. unknown
Dr. Smith's Automaton (Fr.), dir. unknown
Frankenstein, dir. J. Searle Dawley*

---

* Only the major *Frankenstein* films are listed here. For a more extensive listing see Walt Lee, *Reference Guide to Fantastic Films*, Los Angeles: Chelsen-Lee Books, 1972, I, pp. 148–150.

The Mechanical Husband, dir. unknown
Mechanical Mary Anne (Br.), dir. Lewin Fitzhamon

1911   *The Automatic Motorist* (Br.), dir. Walter Booth
The House of Mystery (Fr.), dir. unknown
How They Work in the Cinema, dir. unknown
The Tyrolese Doll (Fr.), dir. unknown
Automaton or Acrobat? (Fr.), dir. unknown

1912   *Coppélia* (Danish), dir. unknown

1913   *The Doll and the Devil* (Pilot film production), dir. unknown

1914   *The Electric Doll*, dir. unknown
The Golem (Ger.), dir. Paul Wegener
The Tales of Hoffmann (Ger.), dir. Richard Oswald

1915   *The Dancing Doll*, dir. unknown
Life Without Soul, dir. Joseph W. Smiley
The Mechanical Man, dir. Curtis

1916   *The Golem* (Danish), dir. Urban Gad
Homunculus (Ger.), dir. Otto Rippert (serial in 6 episodes)

1917 *The Golem and the Dancing Girl* (Ger.), dir. Paul Wegener

1918 *Alraune* (Ger.), dir. unknown

1919 *The Doll* (Ger.), dir. Ernst Lubitsch
The Master Mystery, dir. Burton King (serial in 15 episodes)

1920 *Algol* (Ger.), dir. Hans Werckmeister
The Golem: How He Came Into the World (Ger.), dir. Paul Wegener
La Poupée (Br.), dir. Meyrick Milton

1921 *The Golem's Last Adventure* (Austrian), dir. Julius Szomogyi

1923 *The Miracle of Tomorrow* (Ger.), dir. Harry Piel

1924 *Aelita* (USSR), dir. I Protozanov
Thief of Bagdad, dir. Raoul Walsh
Waxworks (Ger.), dir. Paul Leni
A Machine that Thinks, dir. unknown

1926 *Chess Player* (Fr.), dir. Raymond Bernard
Metropolis (Ger.), dir. Fritz Lang

1928 *Alraune* (Ger.), dir. Henrik Galeen
The March of the Machines (Belgian), dir. Eugene Deslaw

1929 *Mysterious Island*, dir. Lucien Hubbard

1930 *Alraune* (Ger.), dir. Richard Oswald

1931 *Frankenstein*, dir. James Whale

1932 *Robot*, dir. unknown
(Max Fleischer cartoon)
Robots (Fr.), dir. unknown

1933 *Love Has Its Reasons* (Ger.), dir. Hans Steinhoff

1934 *March of the Wooden Soldiers* (*Babes in Toyland*), dir. Gus Meins
Master of the World (Ger.), dir. Harry Piel
The Vanishing Shadow, dir. Lew Landers (serial in 12 episodes)

1935 *Birth of a Robot* (Br.), dir. Len Lye; music by Gustav Holst
Bride of Frankenstein, dir. James Whale
The Golem (Fr. & Czech), dir. Julien Duvivier
The Phantom Empire, dir. B. Reeves Eason and Otto Brower (serial in 12 episodes)
Loss of Sensation (USSR), dir. ?
The Tin Man (Hal Roach short for M.G.M.)

1936 *The Devil Doll*, dir. Tod Browning
The Undersea Kingdom, dir. B. Reeves Eason and Joseph Kane (serial in 12 episodes)

1937 *Mechanical Handy Man*, dir. ?

1938 *The Chess Player* (Fr.), dir. Jean Dréville

1939 *Buck Rogers*, dir. Ford Beebe and Saul Goodkind (serial in 12 episodes)
The Phantom Creeps, dir. Saul Goodkind and Ford Beebe (serial in 12 episodes)
Pinocchio, dir. Walt Disney
The Wizard of Oz, dir. Victor Fleming (Tin Man and living scarecrow)
Zorro's Fighting Legion, dir. William Witney and John English (serial)

1940 *Flash Gordon Conquers the Universe*, dir. Ray Taylor (serial in 12 episodes)
Son of Frankenstein, dir. Rowland V. Lee

The *Mysterious Dr. Satan* (*Dr. Satan's Robot*), dir. William Whitney (serial in 15 episodes)
Thief of Bagdad (Br.), dir. Ludwig Berger, Tim Whelan, and Michael Powell (mechanical horse)

1941 *Cracked Nuts*, dir. Eddie Cline
Hellzapoppin, dir. H. C. Potter (brief appearance of Frankenstein's monster)
Robot Wrecks, dir. Edward Cahn ("Our Gang" comedy)

1942 *The Boogieman Will Get You*, dir. Lew Landers
The Ghost of Frankenstein, dir. Erle C. Kenton

1943 *Frankenstein Meets the Wolfman*, dir. Roy Neill

1944 *The House of Frankenstein*, dir. Erle C. Kenton

1945 *Dead of Night* (Br.), dir. Alberto Cavalcanti (episode of "live" ventriloquist dummy)
The Monster and the Ape, dir. Howard Bretherton (serial in 15 episodes)

1948 *Abbott and Costello Meet Frankenstein*, dir. Charles Barton
Federal Agents vs. Underworld, Inc., dir. Fred Brannon (serial in 12 episodes)
One Touch of Venus, dir. William A. Seiter

1949 *The Perfect Woman* (Br.), dir. Bernard Knowles

1950 *Dopey Dicks*, dir. Hugh McCollum
Magic Sword (Yug.), dir. Voja Nanovic

1951 *Captain Video*, dir. Spencer G. Bennett

and Wallace A. Grissell (serial in 15 episodes)
The Day the Earth Stood Still, dir. Robert Wise
The Emperor and the Golem (Czech), dir. Martin Fric
Tales of Hoffmann (Br.), dir. Michael Powell and Emeric Pressburger

1952 *Alraune* (Ger.), dir. Arthur-Maria Rabenalt
My Son the Vampire (*Old Mother Riley Meets the Vampire*) (Br.), dir. John Gilling (despite its title, the film deals with robots and other sci-fi elements)
Twisted Neck vs. Frankensberg (Fr.), dir. Paul Paviot
Zombies of the Stratosphere, dir. Fred Brannon (serial in 12 episodes)

1953 *The Four-Sided Triangle* (Br.), dir. Terence Fisher
Magnetic Monster, dir. Curt Siodmak
Robot Monster, dir. Phil Tucker
The Twonky, dir. Arch Oboler
The War of the Worlds, dir. Byron Haskin

1954 *Gog*, dir. Herbert L. Strock
Rocky Jones Space Ranger (TV series) launched
Target Earth, dir. Sherman A. Rose
Three Were Three (Span.), dir. Eduard Maroto
Tobor the Great, dir. Lee Sholem
Robot Rabbit, dir. Fritz Freleng (short cartoon)

1955 *Bowery Boys Meet the Monsters*, dir. Edward Bernds
Creature with the Atom Brain, dir. Edward L. Cahn
The Flying Saucers (Mexican), dir. Julian Soler

1956 *Bride of the Monster*, dir. Edward Wood
Earth Versus the Flying Saucers, dir. Fred F. Sears

Forbidden Planet, dir. Fred Wilcox
Invasion of the Body Snatchers, dir. Don Siegel (deals with cloning or reduplication)

1957   *The Curse of Frankenstein* (Br.), dir. Terence Fisher
I Was a Teenage Frankenstein, dir. Herbert L. Strock
The Invisible Boy, dir. Herman Hoffman
Kronos, dir. Kurt Neuman
The Mysterians (Japanese), dir. Inoshira Honda
Twenty Faces (Japanese), dir. Hideo Sekigawa (robot beetles)
The Twenty-Seventh Day, dir. William Asher

1958   *Attack of the Puppet People*, dir. Bert I. Gordon
The Colossus of New York, dir. Eugene Lourie
First Spaceship on Venus (Polish–E. German), dir. Kurt Maetzig
Frankenstein 1970, dir. Howard W. Koch
Frankenstein's Daughter, dir. Richard Cunha
How to Make a Monster, dir. Herbert L. Strock

1959   *Have Rocket Will Travel*, dir. David Rich
The Robot vs. the Aztec Mummy (Mexican), dir. Rafael Portillo

1960   *Beauty and the Robot*, dir. Albert Zugsmith
Robot (Yug.), dir. Sasa Dobrila

1961   *Goliath and the Vampires* (It.), dir. Sergio Corbucci and Giacomo Gentilomo
Hercules in the Haunted World (It.), dir. Mario Bava (giant "living" statue)
Maszyna (Polish), dir. unknown
Most Dangerous Man Alive, dir. Alan Dwan

Neutron vs. the Death Robots (Mexican), dir. Federico Curiel
Venera (Yug.), dir. ?

1962   *Attack of the Robots* (Fr.–Span.), dir. J. Franco
The Brain that Wouldn't Die, dir. Rex Carlton
Citizen IM5 (Yug.), dir. ?
Creation of the Humanoids, dir. Wesley E. Barry
Rocket to Nowhere (Czech), dir. Jindrich Polak
Storm Planet (USSR), dir. Pavel Klushantsev
Portrait — Robot (Fr.), dir. ?
Voyage to a Prehistoric Planet, dir. John Sebastian (U.S. version of *Storm Planet* (1962))
Year 3003 (Yug.), dir. ?

1963   *Devil Doll*, dir. L. Shonteff
Friend (Polish), dir. unknown
Hypnosis (Span. – It.), dir. Eugenio Martin ("living" ventriloquist dummy)
Jason and the Argonauts, dir. Don Chaffey (episode of Talos, the giant of bronze)
Outer Limits (TV series) launched (memorable robot episodes: "I, Robot," "Demon with a Glass Hand")

1964   *Dr. Orloff's Monster* (Span.), dir. J. Franco
Fantômas (Fr.), dir. André Hunnebelle
Frankenstein Conquers the World (Japanese), dir. Inoshira Honda
Goliath and the Vampires (It.), dir. Sergio Corbucci and Giacomo Gentilomo
Hercules vs. the Giant Warriors (It.), dir. Alberto De Martino
Kiss Me Quick, dir. Russ Meyer (?) (Frankenstein's monster appears along with a vampire and a mummy)
The Earth Dies Screaming (Br.), dir. Terence Fisher
The Evil of Frankenstein (Br.), dir. Freddie Francis
The Human Duplicators, dir. Hugo Grimaldo

The Spy with My Face, dir. John Newland

The Testament of Frankenstein (Span.), dir. Jose Luis Madrid

The Time Travelers, dir. Ib Melchior

1965   *Alphaville* (Fr.), dir. Jean-Luc Godard (Lemmy Caution vs. a super-computer)

Dr. Goldfoot and the Bikini Machine, dir. Norman Taurog

Dr. Who and the Daleks (Br.), dir. Gordon Flemyng

Frankenstein Meets the Spacemonster, dir. Robert Gaffney

Lost in Space (TV series) launched

My Mother, the Car (TV series) launched (car appears as "robot" of reincarnated mother)

Sins of the Fleshapods, dir. Mike Kuchar

1966   *The Amazing Dr. G.* (It.), dir. G. Simonelli

Carry on Screaming (Br.), dir. Gerald Thomas

Castle of Evil, dir. Francis D. Lyon

Cyborg 2087, dir. Franklin Adreon

The Disintegrating Ray, or The Adventures of Quique and Arthur the Robot (Span.), dir. Pascual Cervera

Dr. Coppelius (Span.–U.S.), dir. Ted Kneeland

Dr. Goldfoot and the Girl Bombs (U.S.–It.), dir. Mario Bava

Frankenstein Created Woman (Br.), dir. Terence Fisher

Gulliver's Travels Beyond the Moon (Japanese), dir. Yoshio Kuroda

How to Steal the Crown of England (It.), dir. Terence Hathaway

Jesse James Meets Frankenstein's Daughter, dir. William Beaudine

Majin (Japanese), dir. Kimiyoshi Yasuda (giant armored statue comes to life)

Star Trek (TV series) launched (memorable robot episodes: "I, Mudd," "What Are Little Girls Made Of?" "The Changeling")

Superego vs. the Robots (It.–Span.), dir. Paolo Binachini

The Golem (Fr.), dir. Jean Kerchbron

Time Tunnel (TV series) launched

Venetian Affair, dir. Jerry Thorpe

Water Cyborgs (Japanese), dir. Hapime Sato

1967   *Daleks — Invasion Earth 2150 A.D.* (Br.), dir. Gordon Flemyng

Fearless Frank, dir. Philip Kaufman

It!, dir. Herbert J. Leder

Number 00173 (Polish), dir. unknown

Planet of the Three Islands (Polish), dir. unknown

The Fantastic Three (It.–W. Ger.), dir. G. Parolini (a machine has the power to duplicate people)

The President's Analyst, dir. Theodore J. Flicker

The Terrornauts (Br.), dir. Montgomery Tully

War of the Gargantuas (*The Frankenstein Brothers*) (Japanese), dir. Inoshira Honda

Wild, Wild Planet (It.), dir. Antonio Margheriti

1968   *The Astro-Zombies*, dir. Ted Mikels

Bang, Bang (U.S.–Span.), dir. Stanley Praeger

Barbarella (Fr.–It.), dir. Roger Vadim (devil-dolls)

Journey to the Unknown (TV series) launched

King Kong Escapes (Japanese), dir. Inoshira Honda (the Great Ape fights a monster-robot)

Out of the Unknown (TV series) launched

Prague Nights (Czech), dir. Jiri Brdecka

The Colossus of New York, dir. Eugene Lourie

Torture Garden (Br.), dir. Freddie Francis

2001: A Space Odyssey, dir. Stanley Kubrick (man vs. computer)

1969   *Der Roboter* (E. Ger.), dir. unknown

Frankenstein De Sade, or Hollow my Weenie, Dr. Frankenstein, dir. unknown

Mad Monster Party, dir. Jules Bass

One More Time, dir. Jerry Lewis

Scream and Scream Again (Br.), dir. Gordon Hessler

S.O.S. Invasion (Span.), dir. Silvio F. Balbuena

Some Girls Do (Br.), dir. Ralph Thomas

Swiss Made (Swiss), dir. Yves Yersin

The Computer Wore Tennis Shoes, dir. Robert Butler

The Pleasure Machine, dir. Ron Garcia

The Wrestling Women vs. the Murdering Robot (Mexican), dir. unknown

Troika, dir. Frederic Hobbs

THX 1138, dir. George Lucas

1970   *Colossus, the Forbin Project*, dir. Joseph Sargent

Dracula vs.. Frankenstein (Span.), dir. Tulio Demichel

Dr. Frankenstein on Campus (Canadian), dir. Gilbert W. Taylor

Electro-Sex, dir. unknown

Flick (Canadian), dir. Gilbert W. Taylor

Frankenstein must be Destroyed (Br.), dir. Terence Fisher

Necropolis (It.), dir. Franco Brocani

Night Gallery (TV series) launched

Robot Agent X-2, dir. unknown

The Curious Female, dir. Paul Rapp

The Girl of Tin (It.), dir. Marcello Aliprandi

The Horror of Frankenstein (Br.), dir. Jimmy Sangster

Thirty Thousand Miles under the Sea (Japan), dir. unknown

1971   *Dracula vs. Dr. Frankenstein* (Span.), dir. J. Franco (reference sources may have confused this with Tulio De-

michel's film listed under 1970)

Egghead's Robot, dir. Milo Lewis

Escape, dir. John L. Moxey

Lady Frankenstein (It.), dir. Mel Wells

1972   *Frankenstein and the Monster from Hell* (Br.), dir. Terence Fisher

The Questor Tapes (TV series) launched

Silent Running, dir. Douglas Trumbull

1973   *The Six Million Dollar Man* (TV series) launched

Sleeper, dir. Woody Allen

Solaris (USSR), dir. Andrei Tarkovsky

Westworld, dir. Michael Crichton

1974   *Andy Warhol's Frankenstein in 3D* (*Flesh for Frankenstein*), dir. Paul Morrissey

Fantastic Planet (Fr.–Czech), dir. Rene Laloux

The Stepford Wives, dir. Bryan Forbes

Young Frankenstein, dir. Mel Brooks

1975   *Flesh Gordon*, dir. Michael Benveniste and Howard Ziehm

1976   *Fellini's Casanova* (It.), dir. Federico Fellini (life-size female robot; animated bird)

Futureworld, dir. Richard T. Heffron

Logan's Run, dir. Michael Anderson

Demon Seed, dir. Donald Hammell

1977   *Star Wars*, dir. George Lucas